U.S. Marine Corps Combat Water Survival

Plus

U.S. Marine Corps Afghanistan: An Introduction to the Country and People

Department of Defense

Pentagon
Publishing

MCRP 3-02C

Marine Combat Water Survival

U.S. Marine Corps

PCN 144 000069 00

To Our Readers

Changes: Readers of this publication are encouraged to submit suggestions and changes that will improve it. Recommendations may be sent directly to Commanding General, Marine Corps Combat Development Command, Doctrine Division (C 42), 3300 Russell Road, Suite 318A, Quantico, VA 22134-5021 or by fax to 703-784-2917 (DSN 278-2917) or by E-mail to **morgann@mccdc.usmc.mil**. Recommendations should include the following information:

- Location of change
  - Publication number and title
  - Current page number
  - Paragraph number (if applicable)
  - Line number
  - Figure or table number (if applicable)
- Nature of change
  - Add, delete
  - Proposed new text, preferably double-spaced and typewritten
- Justification and/or source of change

Additional copies: A printed copy of this publication may be obtained from Marine Corps Logistics Base, Albany, GA 31704-5001, by following the instructions in MCBul 5600, *Marine Corps Doctrinal Publications Status.* An electronic copy may be obtained from the Doctrine Division, MCCDC, world wide web home page which is found at the following universal reference locator: **http://www.doctrine.usmc.mil**.

Unless otherwise stated, whenever the masculine gender is used, both men and women are included.

DEPARTMENT OF THE NAVY
Headquarters United States Marine Corps
Washington, D.C. 20380-1775

6 January 2003

FOREWORD

Marine Corps Reference Publication (MCRP) 3-02C, *Marine Combat Water Survival*, provides Marine Corps combat water survival techniques, procedures, and training standards. This publication also teaches Marines to cross water obstacles and perform water rescues correctly and safely.

This publication is the foundation for teaching Marines correct water survival techniques and procedures that are used throughout the Marine combat water survival program (MCWSP). Once an individual or a unit has completed the MCWSP, this publication can be used as a refresher course before water operations.

MCRP 3-02C supersedes Fleet Marine Force Reference Publication (FMFRP) 0-13, *Marine Combat Water Survival*, dated 16 September 1991.

Reviewed and approved this date.

BY DIRECTION OF THE COMMANDANT OF THE MARINE CORPS

EDWARD HANLON, JR.
Lieutenant General, U.S. Marine Corps
Commanding General
Marine Corps Combat Development Command

Publication Control Number: 144 000069 00

MCRP 3-02C Marine Combat Water Survival

Table of Contents

Chapter 1. Survival at Sea

Chapter 2. Water Rescues

Chapter 3. Treatment of Casualties and Avoidance of Dangerous Marine Life

Chapter 4. Negotiating Water Obstacles

Chapter 5. Fording Waterways

Appendix. Knot Tying

Throughout history, water has posed special challenges to Marines and Sailors during times of both peace and war. Therefore, the inherent dangers associated with waterborne operations demand that Marines and Sailors receive proper water survival training. Combat units that are confident in their ability to work in and around water can use the water to their advantage in combat, and history is filled with examples of how the proper preparation or training for survival in water reduced or averted disaster.

USS *Indianapolis*

On Sunday, 29 July 29 1945, the heavy cruiser USS *Indianapolis* (the *Indy*) was en route to the Philippine Sea. Shortly before midnight (about 39 hours out of port), the *Indy*, running blacked out and unescorted, was rocked by two explosions on her starboard side. With communications smashed, the ship could not signal its distress and sank within 15 minutes.

Three life rafts and a floater net supported a few survivors, but the rest of the survivors drifted about, held up by rubber life belts or Mae Wests. About 60 seamen died the first night. Survivors assumed the ship would be reported overdue in Leyte, and that they would be rescued within 2 days. By Monday evening, however, panic began to set in as some life jackets lost their buoyancy from the long immersion. Some men even fought over life jackets, which resulted in at least 25 deaths. No one dared sleep for fear of losing his jacket.

Throughout the next several days, in-transit aircraft flew nearby without spotting the desperate seamen. As best they could, the men kept together, some tying long ropes to each other, floating like corks on a net.

Not until late Thursday morning, 3 1/2 days after the ship sank, were the men discovered. Luckily, a plane on a routine flight over the area spotted the survivors. When ships picked them up that night, the survivors learned they had never been reported overdue. Every one of the 1,196-man crew was a casualty; 880 crewmembers were listed as dead or missing. Although many lives were lost, the innovative and expedient use of flotation devices and float techniques employed by the survivors helped save hundreds of lives.

USS *America*
On Thanksgiving eve, 23 November 1995, the USS *America* made its way through the Arabian Sea. Twenty-year-old, Marine Lance Corporal (LCpl) Zachary Mayo was unable to sleep and, wanting some fresh air, made his way onto an open-air platform near the aircraft hangar bay, which was three levels below the sleeping quarters. While he was on the platform, the ship veered suddenly, throwing LCpl Mayo through the platform's protective bars and into the sea, 30 feet below.

Frantic, LCpl Mayo called out in vain to the watchmen on the flight deck, which was 64 feet above him. It soon became clear to him that the USS *America* would keep its course into the Gulf of Oman until his absence was discovered at morning muster. The LCpl took a moment to consider his situation. Since land was at least 100 miles away, swimming was suicide; he would have to stay afloat until a search party found him.

Using the techniques he had learned during combat water survival training, LCpl Mayo made a flotation device out of his coveralls and tried to relax. Meanwhile, business continued as usual aboard the USS *America*. Since LCpl Mayo was on special assignment with the hazardous materials division, his absence went unnoticed until a petty officer asked several Sailors if they had seen their shipmate recently. By the time a roll call had been completed, LCpl Mayo had been adrift at sea for over 24 hours. Although three, fixed-wing Viking aircraft were deployed to search for LCpl Mayo, most people aboard ship feared the worst.

After 34 hours at sea, LCpl Mayo was discovered by fishermen on a Pakistani fishing boat. LCpl Mayo's survival is a testament not only to his incredible physical courage, but also to the soundness of the lifesaving training and techniques he received during combat water survival training.

Chapter 1
Survival at Sea

As a Marine, you face a variety of potential water emergencies whenever you cross expanses of water: ships, watercraft, and amphibious assault vehicles (AAVs) can sink; aircraft can crash into the sea; or you can accidentally fall into the water. However, there are some basic precautionary measures you can take to protect your safety and reduce your chances of becoming a water casualty. Determine the following information as soon as you board any type of vessel. Your knowing the following information may save your life or the life of your fellow shipmates.

- How many life preservers and lifeboats/rafts are on board?
- Where are the life preservers and lifeboats/rafts located?
- What type of unit survival equipment is on board?
- Are individual survival kits issued to each person on board?
- How much food, water, and medicine do the survival kits contain? When was the last time the contents were inspected for proper quantities and shelf life expiration?
- Is there sufficient survival equipment available for the number of personnel?
- How many other personnel are there on board, and where are they located?
- What are the egress procedures for the ship, boat, watercraft, AAV, or aircraft?

Abandoning Ship

When you embark on a Navy ship, you will receive abandoning ship instructions from Navy personnel. If given the order to abandon ship, report to your designated assembly area and put on a life preserver. DO NOT inflate the life preserver until you are clear of the ship. Torn life preservers will not inflate and inflated life preservers can block you, and those behind you, from exiting the ship. A flotation device that has been inflated may also burst if you jump from a significant height. See pages 1-15 through 1-35 for staying afloat with and without a life preserver.

DO NOT remove your clothing, boots, or shoes before abandoning ship. Your trousers and blouse may be the only flotation devices available if your life preserver is faulty or becomes damaged, and your clothes can provide some insulation from the cold water. However, remove your soft cover and place it in a cargo pocket for later use. The soft cover is both lightweight and a good protection against sunburn caused by the sun's rays reflecting off the water.

Jettisoning Equipment

Equipment should be kept properly packed and waterproofed in case you have to abandon ship. If entering the water from a height greater than 30 feet, wearing your equipment (e.g., pack, helmet, gas mask) could cause injury. Upon impact with the water, the helmet will "cup" air inside of it. The chin strap may also create a "hanging effect" as you submerge from the force of the fall. Therefore, you should remove your helmet and gas mask before abandoning ship.

If you are unable to maintain buoyancy due to the amount of equipment secured to your pack and body, then jettisoning some of your equipment may become necessary. Equipment that you should always retain includes canteens of freshwater, first aid kit, soft cover, and survival kit. The survival kit should include first aid items, water purification tablets or drops, fire starting equipment, signalling items (e.g., flashlight, strobe light, chemlights), food procurement items, and shelter items. Other items in a survival kit include sunburn lotion and lip balm, knife, goggles/sunglasses, plastic bag, matches and lighter, mirror. See Marine Corps Reference Publication (MCRP) 3-02F, *Survival*, for a detailed list of survival items and applications.

Abandoning Ship Technique

When abandoning ship, safety considerations must be observed. Use the following technique when abandoning ship without your combat gear:

Place your hands on their opposite shoulders, forming a criss-cross pattern.

Step to the edge of the ship's deck and check the water below for debris or survivors. If the water is clear, look straight ahead and prepare to jump. If the water is not clear, move to another location.

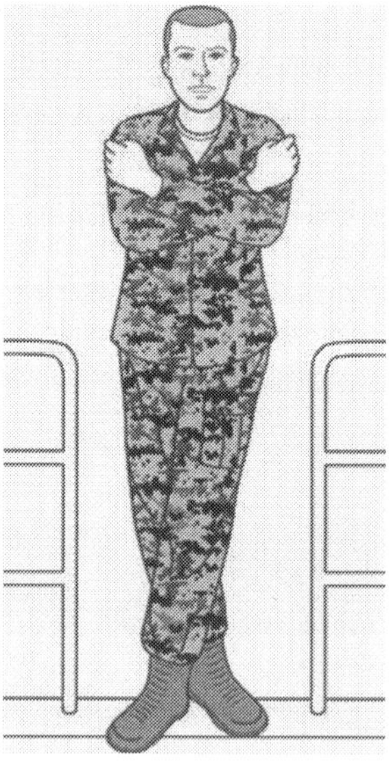

Note: DO NOT hold your nose as you abandon ship. If you do hold your nose, the force of impact into the water could jar your arm and hand and cause you to break your nose.

Step off the side of the ship with a smooth, 30-inch stride. DO NOT DIVE OFF THE SHIP. DO NOT LOOK DOWN AT THE WATER. LOOK STRAIGHT AHEAD. Looking down at the water can render you unconscious or cause injuries upon impact.

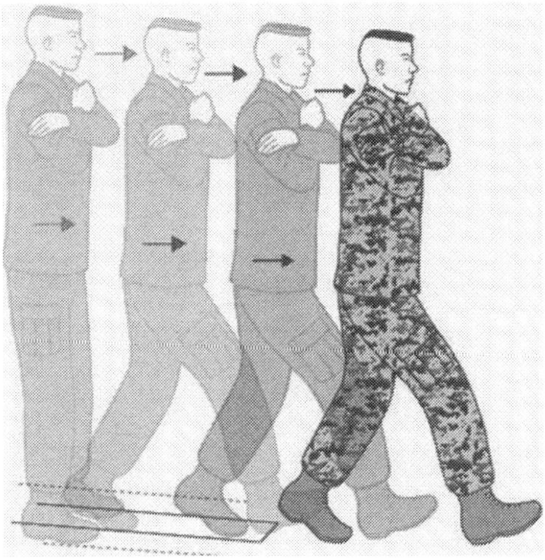

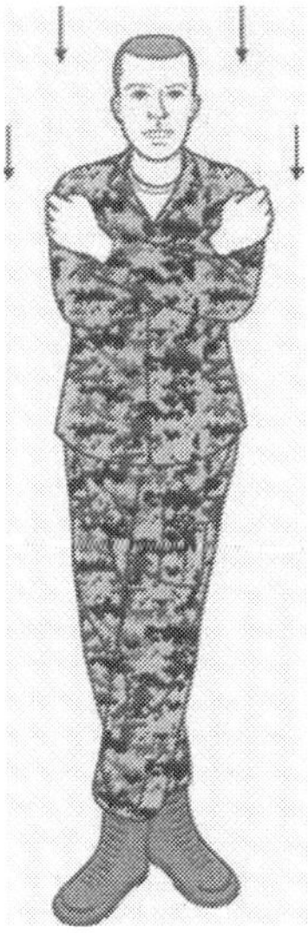

Bring your trailing leg forward during the fall. Cross your trailing leg behind your leading leg.

Keep your head parallel to the water's surface until hitting the water.

You should remain in the abandon ship position until your descent into the water has almost stopped. However, the weight imbalances in your body may cause you to be in a "J" shape under the water

.

Once your downward motion has ceased, your feet may be parallel with the ocean bottom or you may be nearly inverted with your feet over your head. To counteract potential disorientation, you should pause briefly and allow the natural buoyancy of your torso to bring your body to a nearly upright position.

Floating debris can cause hazards. Therefore, you should swim upward, extending one arm (hand is shaped as a fist) upward to feel for obstructions. If you encounter debris, try to push it away or surface in a different location.

Swim away from the ship. DO NOT LOOK BACK AT THE SHIP. Looking back could slow your movement away from the area. Remember, your objective is to leave the area as quickly as possible because—

- Equipment and debris may be falling from or spilling out of the ship.
- Additional casualties can occur if individuals abandoning the ship fall on top of swimmers already in the water.
- Swimmers close to the sinking ship may get pulled underneath the water by the suctioning effect of the ship as it goes under.

Modified Abandoning Ship Technique

When abandoning ship while wearing full combat gear (weapon, helmet, and a properly waterproofed pack), safety considerations must be observed. The modified abandoning ship technique is used while wearing full combat gear and exiting from a height less than 30 feet.

WARNING

When exiting from a height greater than 30 feet, remove your helmet and pack. Fasten the helmet to the pack or place it inside the pack before jettisoning. If jettisoning gear from the ship or aircraft, check the water below for survivors before throwing the gear forward of the intended jump area. Once in the water, retrieve your gear and swim out of the area.

Place your weapon over one shoulder, muzzle down, with the weapon parallel to your side.

Place your arm and hand along the weapon and hold it to your side.

Take your free hand and place it on top of your helmet to prevent neck/spinal injury from the force of the water pulling upward on the helmet as your body enters the water.

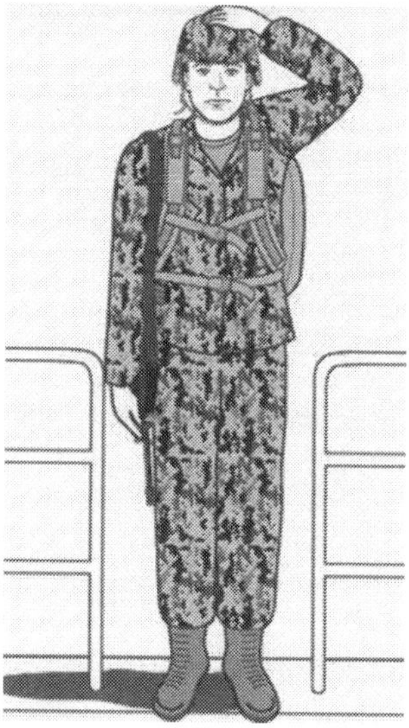

NOTE: An alternate method is to place your weapon over one shoulder, muzzle down, with the weapon parallel to your side. Reach across your body and grasp the sling of your weapon and hold it to your body. Take your free hand and place it on top of your helmet.

Step to the edge of the platform and check the water below for debris and survivors. If the water is clear, look straight ahead and prepare to jump. If the water is not clear, move to another location.

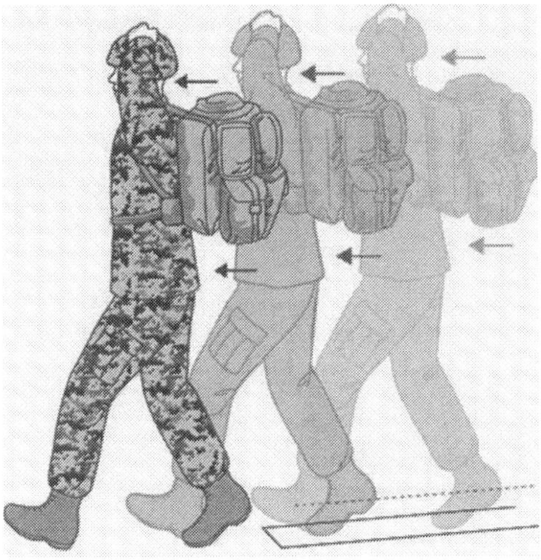

Step off the side of the platform with a smooth, 30-inch stride. DO NOT DIVE OFF THE SHIP, DO NOT LOOK DOWN AT THE WATER, LOOK STRAIGHT AHEAD. Looking down at the water can render you unconscious or cause injury upon impact.

Bring your trailing leg forward during the fall. Cross your trailing leg behind your leading leg.

Keep your head parallel to the water's surface until hitting the water.

You should remain in the modified abandoning ship position until your descent into the water has almost stopped. The buoyancy of a properly waterproofed pack will immediately pull you to the surface. Once you break the water's surface, unsling your weapon and loop the sling over your head. You may have to seesaw the sling so as to ensure the sling passes between your helmet and pack. Once over your head, loop the sling around your neck with the weapon aligned in the center of your body, muzzle down. Lean back on your pack and perform the combat travel stroke to exit the area.

> *NOTE:* You may also remove the pack (unsnap the quick-release strap on one side of the pack, and pass the strap to your other hand in order to maintain contact with the pack, which is functioning as your flotation device). Remain horizontal in the water with your pack under your chest. Perform a breast stroke kick.

Surface Burning Oil Swim

After you have abandoned ship, rise to the surface using the techniques shown on the page 1-6. However, you must remember that fuel from sinking ships or downed aircraft will float on the surface of the water. Therefore, you must move clear of the floating fuel by swimming away from the ship or aircraft as soon as possible. Either swim upwind (into the wind) of the ship/aircraft or swim against the current. Either method allows you to move away from the fuel and the wind/current will push the fuel past you. To properly execute a surface burning oil swim—

Extend your arms overhead as far as possible.

Wave your arms back and forth vigorously to splash a hole while moving upward.

Splash as long as possible to push burning fuel away from the surfacing area.

Use your arms and hands to sweep away fuel and debris.

Kick your legs in a constant breast stroke kick.

Extend your arms (palms outward) forward on the surface, arms shoulder-width apart.

Pull your hands in and back toward the chest.

Stop your hands in front of your face and rotate them so that your palms face forward (roughly halfway out of the water).

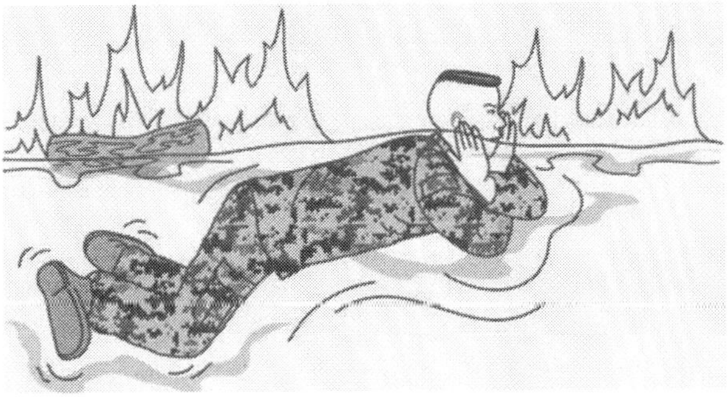

Sweep your arms forward to a full extension at the shoulder width. This splashes debris, oil, or burning liquids aside. To reduce the chance of fatigue, use two short splashes to the front to extend the path.

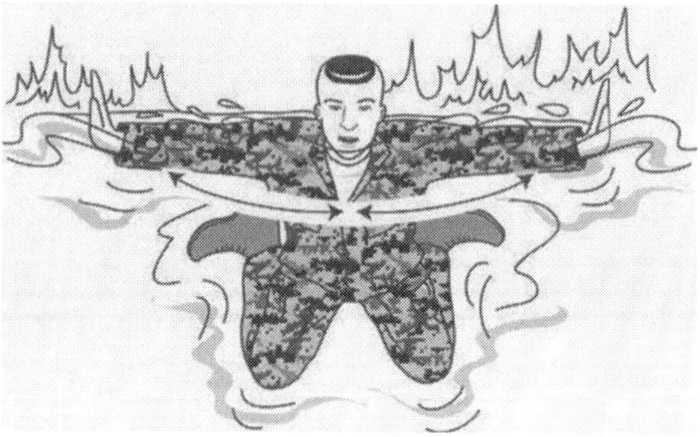

Repeat the preceding step as necessary while swimming clear of the area.

Surviving With a Pack

If packed properly, your pack will float, and it is your key piece of equipment for staying afloat and overcoming water obstacles. If the pack's contents are properly waterproofed, it can support you (with a combat load) in the water. Buoyed up by a waterproofed pack, you eventually emerge from the water with all your equipment (e.g., boots, helmet, flak jacket, weapon, survival items).

Your pack floats based on a scientific principle known as Archimedes' principle. This principle states that an object submerged in a liquid is buoyed up by a force equal to the weight of the liquid displaced (pushed aside) by the object. If the weight of the displaced liquid is greater than the weight of the object, the object floats. If the weight of the displaced liquid is less than the weight of the object, the object sinks. For example, a machine gun sinks in the water, but it still weighs less in the water than it does on land. Even though a machine gun sinks, it is still buoyed up by a force equal to the weight of the water it displaces. For this reason, you should not try to hold yourself or your equipment any higher out of the water than they would naturally float; doing so wastes both energy and body heat.

Preparing Equipment

Before packing your pack, you must prepare your gear/equipment. Tape or pad all sharp edges and equipment corners. Ideally, your gear/equipment is placed in plastic bags and the plastic bag is then placed inside the standard-issued, rubberized, waterproof bag. Waterproof bags are not completely water tight; henceforth, the added protection of first wrapping the gear/equipment in a plastic bag, then placing the plastic bag inside the waterproof bag. Large plastic bags (e.g., trash bags) work well for bulky equipment (e.g., sleeping bags, field jackets, shelter halves, gas masks). Small plastic bags work best for small items (e.g., shaving gear).

> *Note:* If a gas mask must be carried outside the pack, cover it with a waterproof bag.

You will need the following items, which are available through the supply system, to prepare your equipment for packing:

- Waterproof bags:
  Pistol bag, plastic size #1 (8" x 18").
  Rifle bag, plastic size #2 (10" x 56").
  Machine gun bag, plastic size #3 (15" x 56").
  Multipurpose bag, cover, plastic size #4 (20" x 84").

- Panama work vest (life preserver).
- Small plastic bag, self sealing (6" x 6" and 12" x 12").
- Riggers tape, olive drab (2 1/2").
- Willie peter bag, waterproof clothing.

Tying Waterproof/Plastic Bags

Try to remove the excess air from the waterproof/plastic bag before securing the bag's opening. If filled with air, the bag can burst if pressed from the outside. Perform the following steps to tie a waterproof/plastic bag:

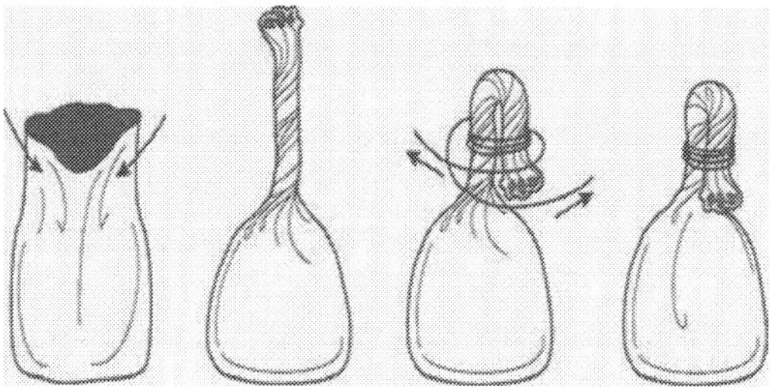

Packing the Pack

Place filled and tied bags inside the pack. Carefully handle sharp items (e.g., tent pegs, poles) to prevent puncturing the bag. Place items in the pack in order of expected use. Close the pack and its compartments. Attach sleeping mats or bags as high as possible on the outside of the pack.

Swimming With the Pack

A useful technique for propelling yourself forward while wearing a pack is the combat travel stroke. This technique allows you to float nearly horizontal and to propel yourself forward with bicycle-style kicks and breast stroke-style arm sweeps. To execute the combat travel stroke—

Body position
Keep the upper part of your body prone to the water, let your legs dangle horizontally, and keep your face up.

Arm action
Extend your hands out in front of your waist. Sweep your arms slightly downward and back 90 degrees to propel your body through the water. Move your hands back to the front of your waist. Repeat.

Leg action
Continuously move your legs in a bicycle-like movement, bringing your knees up high and step out.

Breathing
Keep your face out of the water during the stroke, and breathe freely.

Staying Afloat With a Life Preserver

The best form of flotation is to find any kind of floating object that will keep you and your equipment out of the water or minimize your exposure to the water. Life preservers are the best method, as they allow you to wear your clothes for heat retention and sunburn prevention. Marines use two basic classes of life preservers: inherently buoyant life preservers and inflatable life preservers.

Inherently Buoyant Life Preservers

Inherently buoyant life preservers are either vest-type (worn like a jacket) or yoke-type (worn around the neck). The preserver's outer envelope is either a cotton or water resistant material that encloses a removable fibrous glass or plastic foam filling.

In the Marine Corps, the most common type of inherently buoyant life preserver is the vest-type with collar, known as the kapock preserver. The kapock consists of collar straps, upper front chest straps, leg straps, and waist drawstrings that secure the preserver to you. The leg straps, which are fitted on both sides of the life preserver, ensure that the preserver remains around your chest while you are in the water. A chest strap is attached to the life preserver to facilitate lifting you out of the water. The strap can also be attached to other survivors or to lifeboats to reduce the fatigue that results from holding onto a floating/secured object by hand.

Inflatable Life Preservers

Marine Corps aircraft and AAVs have inflatable life preservers on board. The inflatable life preserver issued to Marines is known as the LPP (life preserver personal). The LPP is capable of both oral inflation and CO_2 cartridge inflation. The LPP consists of buoyancy chambers, CO_2 inflator, and an oral inflation tube. The buoyancy chamber is deck gray in color and is made from a neoprene-coated nylon fabric.

Inflatable life preservers must be stored in a cool, dry place. Heat, moisture, and light cause deterioration of the life preserver material. Do not stow CO_2 cylinders near steam lines or radiators. Heat can increase the pressure inside the cylinders causing them to explode. Avoid sharp edges in stowage. Sharp edges increase wear and tear on the life preservers and may also puncture inflatable buoyancy chambers.

> **CAUTION:** Do not inflate the life preserver until you are clear of the aircraft, ship, or vehicle. Torn life preservers will not inflate and inflated life preservers can block you, and those behind you, from exiting the aircraft, ship, or vehicle.

To don and adjust the LPP—

Remove the LPP from its storage container.

Fasten the belt fasteners in front with the pouch in the rear.

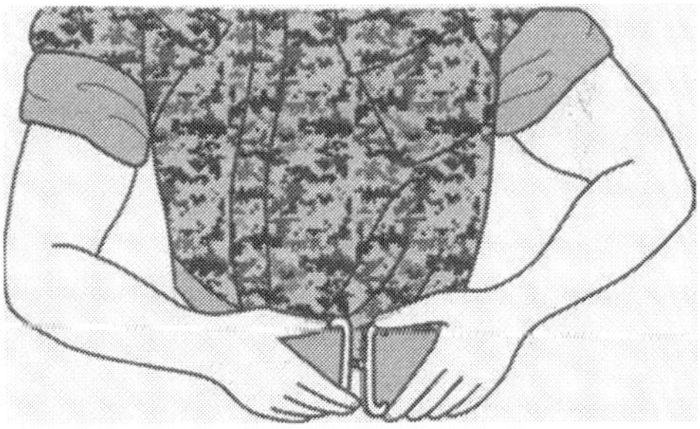

Adjust the belt to fit; secure any excess belt by mating the hook and pile tape.

Rotate the pouch to the front and re-adjust the belt if necessary.

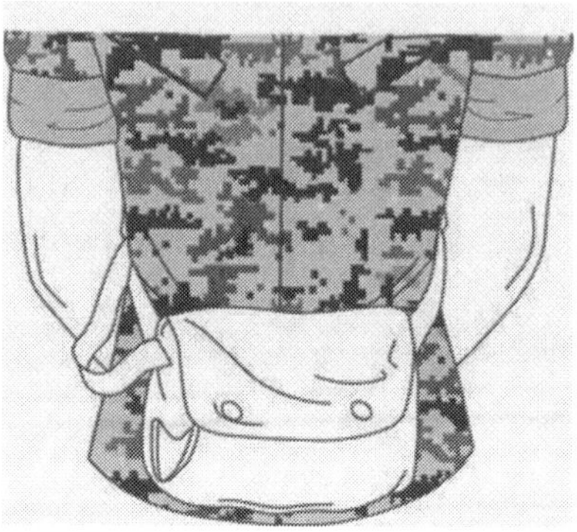

Open the snap fasteners on the pouch and unfold the life preserver.

Place the deflated life preserver over your head.

Place the storage container into the pouch after donning the life preserver.

Lift the lower end of the life preserver out of the pouch.

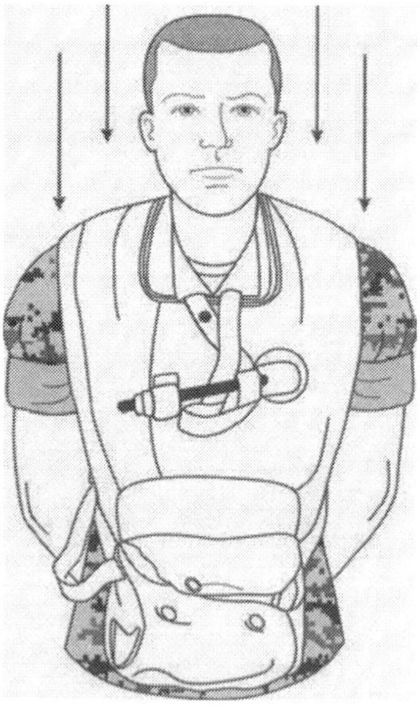

Inflate the life preserver by pulling on the lanyard attached to the CO_2 inflation valve or by blowing on the end of the oral inflation valve.

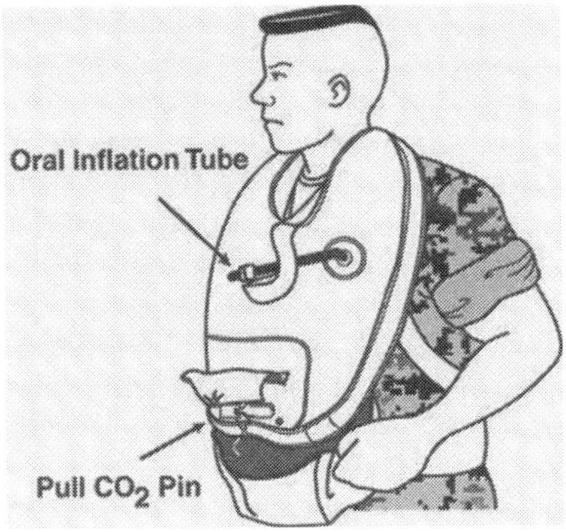

Staying Afloat Without a Life Preserver

You may be in open water without any floating objects or a life preserver to help you survive. If so, uniform blouses and trousers can be made into expedient flotation devices.

Floating With an Inflated Blouse

It is possible to float by a bubble of air trapped in the shoulders of your blouse. The air rises to the back and shoulders of the blouse and supports you at the water's surface. An inflated blouse is also a temporary flotation device used by weaker swimmers while trying to remove their trousers. There is a primary and an alternate way to create a bubble of trapped air in a blouse—

Primary Method
Turn the collar inside the blouse to help create a seal.

Unbutton top button and pull collar around mouth and nose.

Take a deep breath and bend forward slightly at the waist. Exhale one-half to three-quarters of a breath into the blouse.

Grasp and twist the collar with one hand to create a seal, this prevents air from escaping out from the collar.

Use your free hand and feet to stroke and kick to the surface.

Gather and hold the blouse tightly at the collar and stomach level to prevent the blouse from losing air if it floats up too high.

Splash water on the blouse periodically to prevent the material from drying, dry material allows air to escape.

Repeat inflation as required.

Alternate Method

Turn the collar inside the blouse to help create a seal.

Unbutton the second button from the top.

Take a deep breath and bend forward slightly at the waist.

Place your mouth and nose inside the hole created by the open button and exhale one-half to three-quarters of a breath into the blouse.

Grasp material at the unbutton portion and pull downward.

Use your free hand and feet to stroke and kick to the surface.

Splash water on the blouse periodically to prevent the material from drying, dry material allows air to escape.

Repeat inflation as required.

Floating With Inflated Trousers

In warm water, trousers can be used as a primary expedient flotation device. However, in cold water, submerging your head to remove and inflate your trousers results in heat and energy losses that negate the benefit of using the trousers as a flotation device. Once your trousers are inflated, you float motionlessly as if wearing a life preserver. If needed, assume the heat escape lessening posture (known as the HELP position, see page 1-37) to slow heat loss. As trousers dry, air leaks out of the legs. To slow this process, occasionally splash water on the fabric. Reinflate trousers as needed.

Sling Method

The sling method works if you are a strong swimmer or naturally very buoyant. Take the following steps to inflate trousers using the sling method:

Take a deep breath, bend over, and remove your boots.

Note: Retain your boots. Tie the boot laces together and suspend the boots from your blouse or hang them around your neck so that they rest on your chest.

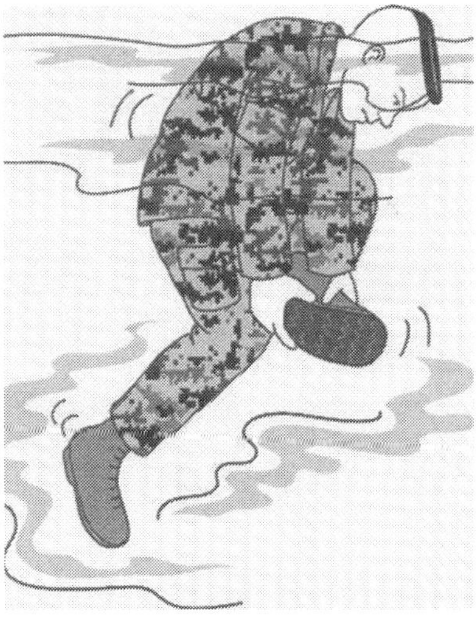

Remove your trousers. Button or zip the trouser's fly closed. This allows you to control airflow.

Tie the bottoms of the trouser legs in a square knot. (The appendix illustrates various knots.)

Ensure that the front (fly) of the trousers faces you.

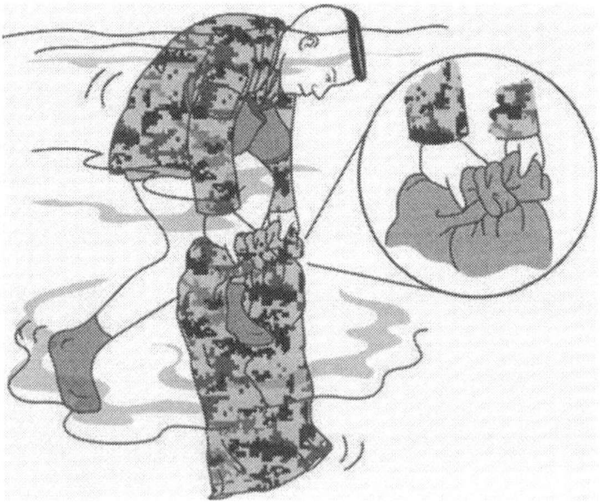

Hold the trousers above the water's surface and behind your head. Grasp both sides of the waistband and open with both hands.

Kick strongly to stay on top of the water while slinging the trousers overhead in order to trap air into them.

Once the waistband is submerged in the water, air is trapped in the legs.

Hold and seal the waistband underwater.

Slip the inflated legs over your head. Hold the waistband in toward your chest, the fly facing your body. To prevent air from escaping from the trousers, seal the waistband by either folding it or twisting it.

Lie back and relax, resting the back of your neck against the knot.

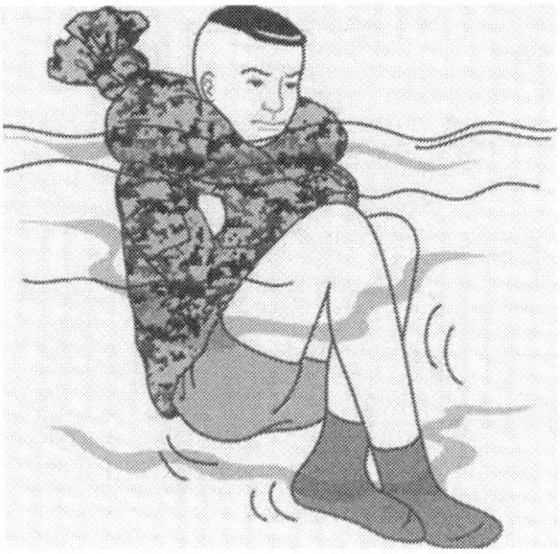

Splash water on the trousers periodically to prevent the material from drying. Dry material allows air to escape.

To replenish air in the trousers, you will use a technique known as the scooping method. With one hand on the open waistband, extend the trousers in front of you just below the surface of the water and scoop air bubbles with your free hand into the open waistband until the trousers have sufficient air. Repeat as necessary.

Splash Method

The splash method is an alternative to the sling method. As with the sling method, you must kick strongly to remain at the surface. To inflate trousers using the splash method, perform the following:

Take a deep breath, bend over, and remove your boots.

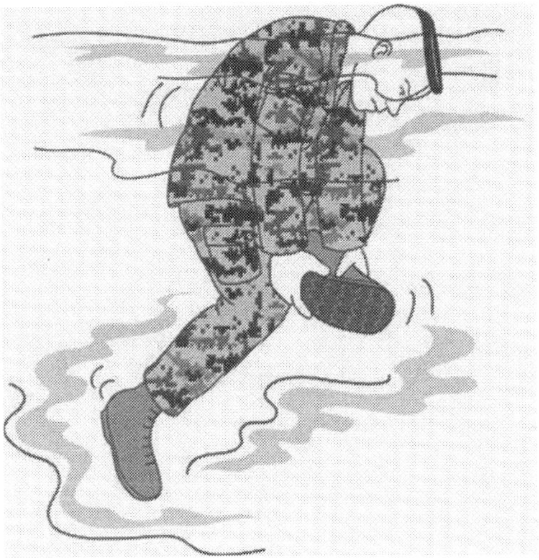

NOTE: Retain your boots. Tie the boot laces together and suspend the boots from your blouse or hang them around your neck so that they rest on your chest.

Remove your trousers. Button or zip the trouser fly closed. This allows you to control airflow.

Tie the bottoms of the trouser legs in a square knot. (The appendix illustrates various knots.)

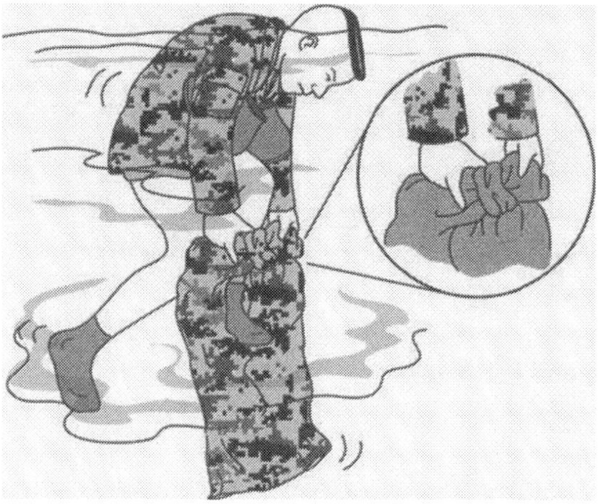

Ensure that the front (fly) of the trousers faces you.

Hold the trousers at the water's surface out in front of you by the waistband with the fly up.

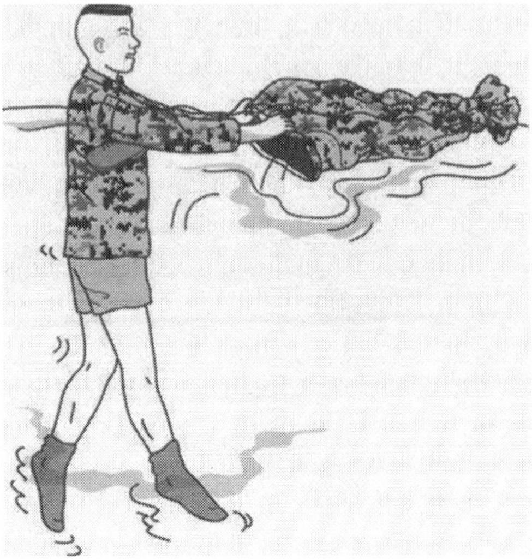

Grasp the waistband at the surface with one hand. Insert your free hand into the waistband, palm down.

Flutter your hand rapidly to create bubbles. This sends a mixture of water and air bubbles into the trousers. The water passes through the fabric. The air remains trapped in the legs.

Hold and seal the waistband underwater.

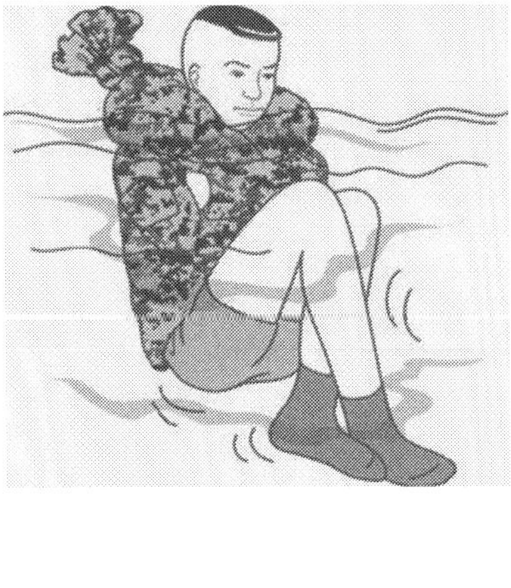

Slip the inflated legs over your head. Hold the waistband in toward your chest, the fly facing your body. To prevent air from escaping from the trousers, seal the waistband by either folding it or twisting it.

Lie back and relax, resting the back of your neck against the knot.

Splash water on the trousers periodically to prevent the material from drying. Dry material allows air to escape.

To replenish air in the trousers, you will use a technique known as the scooping method. With one hand on the open waistband, extend the trousers in front of you just below the surface of the water and scoop air bubbles with your free hand into the open waistband until the trousers have sufficient air. Repeat as necessary.

Blow Method

The blow method is an alternative to the sling method. Use the blow method if you are a weak swimmer. Take the following steps to inflate trousers using the blow method:

Take a deep breath, bend over, and remove your boots.

NOTE: Retain your boots. Tie the boot laces together and suspend the boots from your blouse or hang them around your neck so that they rest on your chest.

Remove your trousers. Button or zip the trouser fly closed. This allows you to control airflow.

Tie the bottoms of the trouser legs in a square knot. (The appendix illustrates various knots.)

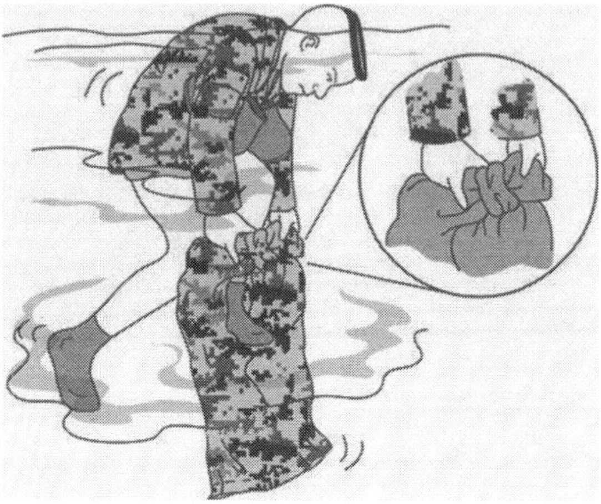

Ensure that the front (fly) of the trousers faces you.

Hold the trousers at the water's surface. Grasp both sides of the waistband and open with both hands.

Take a deep breath.

Drop 2 feet below the water's surface, pulling the waistband underwater.

Hold the waistband open with both hands and blow air into the trousers.

To fill the trousers with air, surface while keeping the waistband underwater, breathe in again, drop below the water's surface, and blow air into the trousers. Repeat these steps until the trousers are filled sufficiently. Once trousers are filled—

Hold the waistband underwater. Twist and pinch it off.

Slip the inflated legs over your head. Hold the waistband in toward your chest, the fly facing your body. To prevent air from escaping from the trousers, seal the waistband by either folding it or twisting it.

Lie back and relax, resting the back of your neck against the knot.

Splash water on the trousers periodically, to prevent the material from drying. Dry material allows air to escape.

To replenish air in the trousers, you will use a technique referred to as the scooping method. With one hand on the open waistband, extend the trousers in front of you just below the surface of the water and scoop air bubbles with your free hand into the open waist-band until the trousers have sufficient air. Repeat as necessary.

Avoiding Heat Loss in Cold Water

The rate of heat exchange in the water is about 25 times greater than it is in air of the same temperature. When you are immersed in cold water, hypothermia occurs rapidly due to the decreased insulating quality of wet clothing and as a result of water displacing the layer of still air that normally surrounds the body. You also lose about 50 percent of your body heat through your head; therefore, keep your head out of the water. Other areas of high heat loss are the neck, the armpits/sides, and the groin.

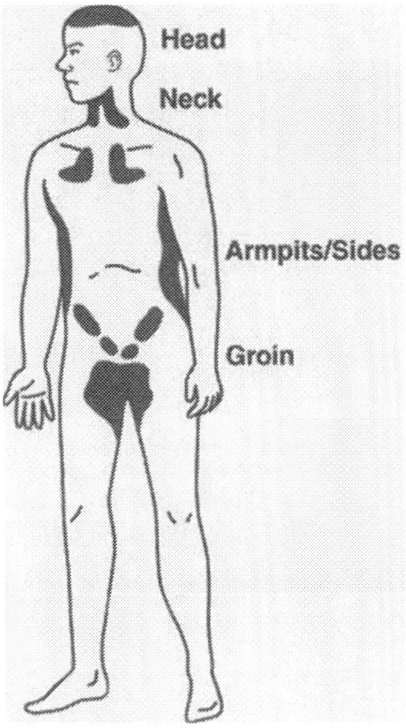

In cold water, DO NOT SWIM TO STAY WARM. Swimming, even with a slow and steady stroke, produces a lot of heat that is lost in the water. The heat loss can produce hypothermia that slows body functions and can result in serious injury or death. Remaining motionless conserves body heat three times longer than swimming. SWIM only if you have a flotation device and the shoreline is visible.

Individual Protection From the Cold

If you are equipped with a life preserver, assume the heat escape lessening posture (known as the HELP position) to slow heat loss and to protect major blood vessels near the body's surface. These areas lack insulating fat and are vulnerable to the chilling effects of cold water. To assume the HELP position—

Tuck your chin down tightly to cover your throat.

Draw your legs up in a fetal position to protect the groin.

Place your arms across your chest, tuck your hands into your armpits.

Wear some type of head covering (e.g., stowed cover, towel, handkerchief) to lessen heat loss through the scalp if head covering is available.

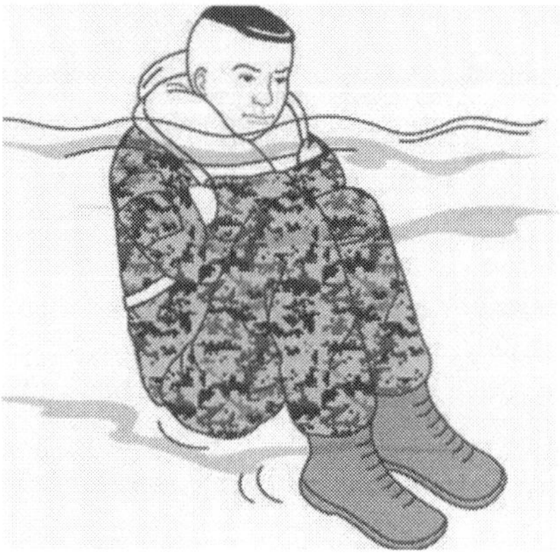

Group Protection From the Cold

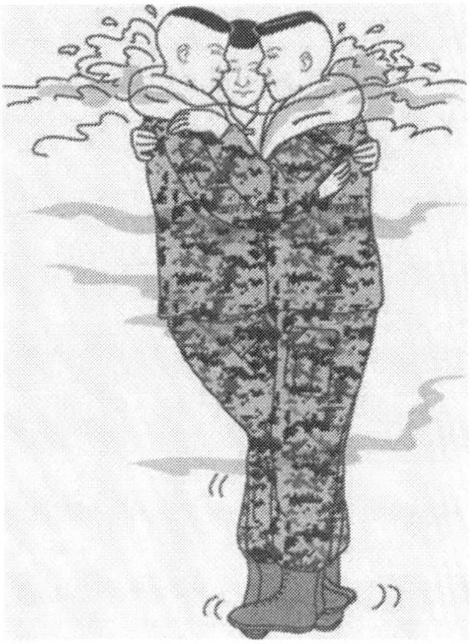

If three or more Marines are in the water and are equipped with life preservers, they should wedge tightly together and lock arms to form a circle known as a huddle position. This position protects vulnerable areas from heat loss.

A casualty who is suffering from the effects of the cold can be placed within the huddle to be surrounded by warmer water. If in the water for a prolonged period, it is recommended that everyone be rotated inside the huddle to maintain or re-warm each person's internal core temperature. If there are more than five Marines, they should make clusters of huddle positions.

Contact with other swimmers provides survival advantages—

Creates a larger target for search and rescue aircraft.

Provides additional warmth in cold water.

Improves morale.

Re-establishes the chain of command.

Reduces shock and panic.

Provides opportunities to administer first aid.

Supports exhausted Marines.

Drownproofing Methods

An object that floats on the surface has positive buoyancy. An object that floats a few feet beneath the surface has neutral buoyancy. An object that sinks has negative buoyancy. Most people have positive buoyancy and will float at the water's surface. To test your buoyancy—

Stand in water that is at your chest level or deeper.

Take a full breath.

Bend forward slowly.

Relax and wait.

If you have positive buoyancy, you will slowly rise to the surface. If you have neutral buoyancy, you will float a few feet beneath the surface. If you have negative buoyancy, you will sink. Regardless of whether you naturally float or sink, you can remain at the surface for extended periods without a life preserver if you exercise the appropriate drownproofing methods or survival strokes, which are based on your buoyancy.

Drownproofing methods consist of the T-method and the sweep. The breast stroke, side stroke, and elementary backstroke are the most common survival strokes. With any drownproofing method or survival stroke, remember the acronym **SAFE**:

SLOW EASY MOVEMENTS
Move slowly to conserve energy and minimize heat loss.

APPLY NATURAL BUOYANCY
Use natural buoyancy to support the body.

FULL LUNG INFLATION
Fill the lungs with each breath. Do not hold air in the cheeks.

EXTREME RELAXATION
Tight muscles are denser than relaxed ones and do not float as well.

Crawl Stroke

The crawl stroke, sometimes called the front crawl or free style, is the fastest stroke. To execute the crawl stroke—

Body position
Lie horizontal, on your stomach, in the water.

Look forward and downward at a 45 degree angle with the waterline between your eyebrows and hairline. Your head position is important as it assists in cutting a path through the

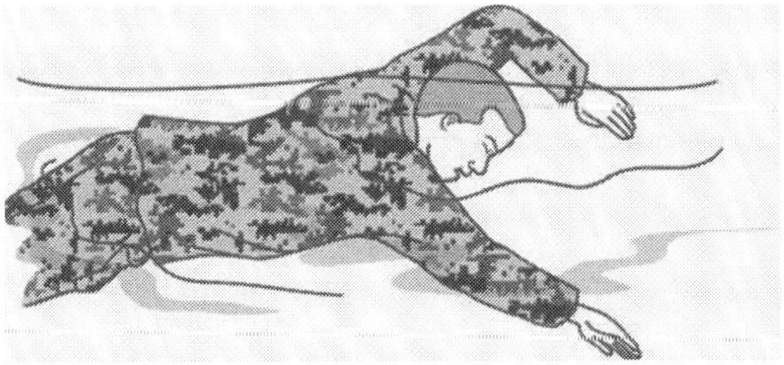

water. If your head is too high in the water, your lower body sinks significantly, making your stroke less efficient. If your head is too low in the water, water washes over your shoulders and neck, causing unnecessary drag.

Arm action
Arm action occurs in three phases: catch, propulsion, and recovery.

Fully extend one arm forward of your body, this positions your hand to catch the water in preparation for the propulsion phase.

To catch the water, bend your wrist (with your palm pointing outboard) and make an "S" shape (or inverted "S" shape) with your hand, ensuring that your hand does not cross the center of your body. Your left hand makes an "S" shape and your right hand makes an inverted "S" shape with your hand finishing at shoulder level.

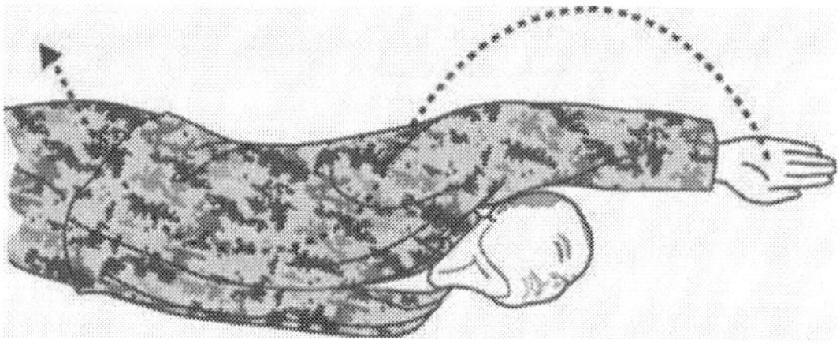

Push with your hands in a rearward fashion toward your feet until your arm is fully extended along your side, keeping your hands close to your body.

To begin the recovery phase, bend your arm at the elbow and raise your hand out of the water. Your hand breaks the surface of the water and maintains a height of 2 to 3 inches above the water's surface.

With your hand and arm moving in a forward manner, bring your hand past your head until your arm is about three-quarters of the way extended. At this point, turn your hand with your palm outboard so as to allow your thumb and fore-finger to enter the water first.

Once your hand enters the water, continue to push your arm forward until it is fully extended.

You are now prepared to catch the water again. These steps are performed in an alternating pattern: when one arm is catching and propelling, the other arm is recovering.

Leg action
Use a flutter kick to create the leg action for the crawl stroke. This kick is used for both propulsion and keeping the lower body horizontal with the water's surface. The flutter kick is an alternating leg action: one leg is kicking in a downward motion while the other leg is recovering to the surface to prepare for the next kick. The size of the kick ranges from 12 to 15 inches and depends on your height.

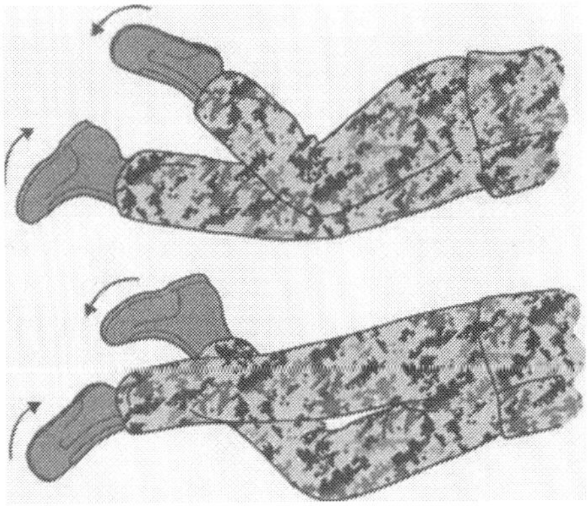

Maintain your legs in a semi-rigid manner.

Generate power for the kick from the hips.

Keep your feet loose at the ankles, they trail behind your legs and act as "flippers."

Execute the propulsion phase of this kick with a downward thrust of your leg.

Execute the recovery phase by pushing your leg back to the surface. This phase is complete when your foot reaches the surface.

Breathing
Breathe during the recovery phase of the arm action.

Roll your body and rotate your head to the side of your body where the arm recovery is occurring, this rolls the water away from your mouth. Keep your chin pushed back toward your shoulder.

Exhale while your face is still submerged and inhale when your face breaks the surface.

Breathe either bilaterally or rhythmically. To breathe rhythmically, breathe on the same side of your body every time your arm cycle occurs. Bilateral breathing requires you to breathe every one and a half arm cycles. To breathe bilaterally, breathe when your right arm is recovering, your face goes back in the water and the next time you breathe is when your left arm is starting its recovery three arm strokes later.

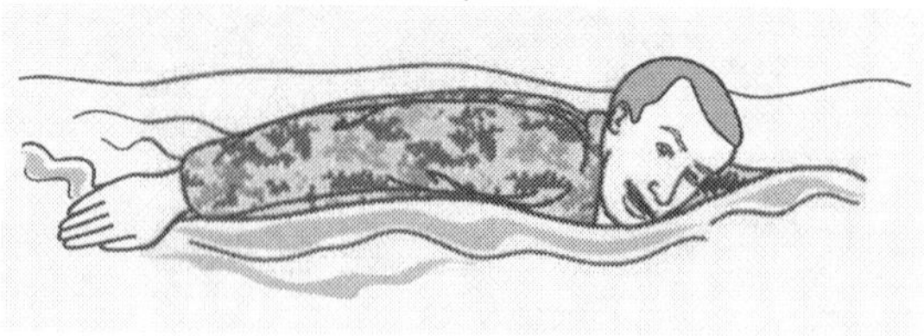

Bilateral breathing is the preferred method for breathing; it prevents the chance of hyperventilation and allows your body to maintain a lateral position in the direction you are swimming.

Coordination
This stroke uses constant arm and leg action.

T-Method

The T-method is a basic drownproofing method. This is the best survival technique if you have negative buoyancy. To execute the T-method—

With your face out of the water, take a deep breath and submerge your face in the water while holding your breath.

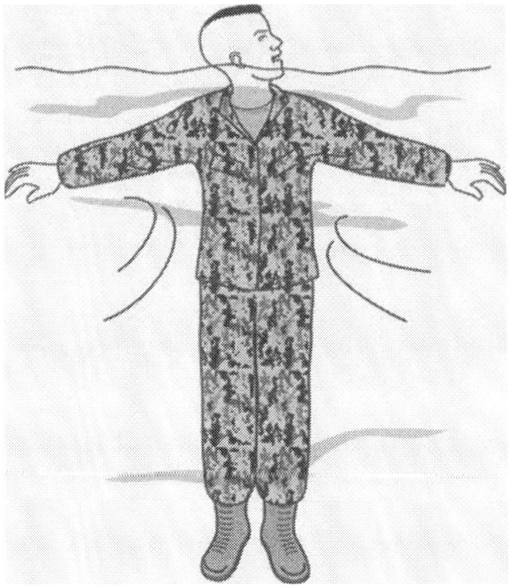

Float with your body in a horizontal position, arms extended from your side, and legs extended and joined.

Move your hands up to your armpits by tracing an imaginary line along your ribs.

Extend your arms outward (horizontal) to your sides, your body position resembles the letter "T."

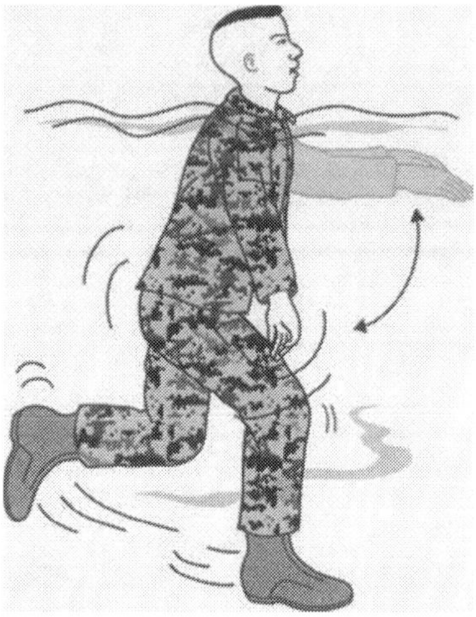

Step out and forward with one leg and point your other leg to the rear, your knees should be slightly bent. Simultaneously, bring your arms down to your sides. Then bring your legs back together. You complete the steps by exhaling most of your air and preparing to surface your face to obtain another breath of air. Hold your head out of the water. Tilt your head back slightly. Breathe normally.

Once your breath is complete, move your hands up and down directly in front of your body. Do this two or three times to slow your descent into the water.

> NOTE: To avoid hyperventilating, hold your breath below the surface of the water for no more than 10 seconds.

The Sweep

The sweep works well if you have slight to excellent positive buoyancy. To execute the sweep—

Float face down in the water: bend 45 degrees at the waist, arms and legs dangling, head hanging down, relax all muscles.

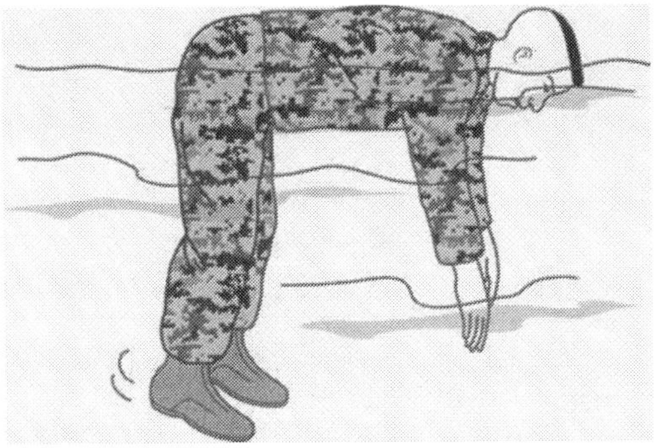

Spread your feet slowly to prepare for a single kick (one leg is forward and one leg is rearward).

Cross your arms in front of your chest, palms outboard with the back of each hand touching the opposite ear.

Exhale prior to raising your head for a breath.

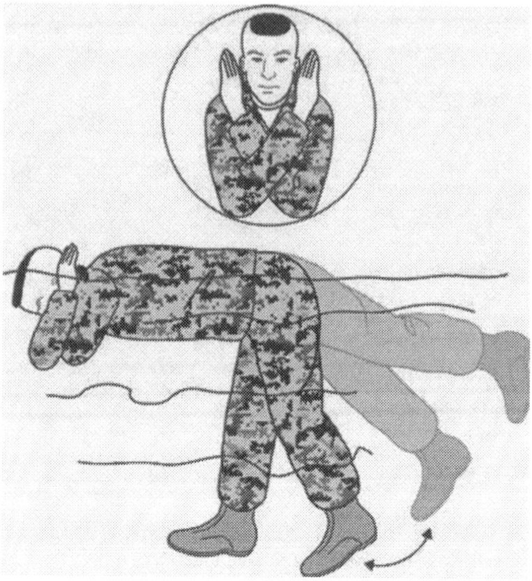

Bring your legs together and sweep your arms down and out until the arms are fully extended out to the sides. This raises your face above the water and allows you to catch a breath of air.

With your air supply replenished, return your face to the water and relax while sweeping your arms in a downward motion in front of your body to prevent/slow your descent. Do not hold your breath for more than 10 seconds. More than 10 seconds enhances your chances of shallow water black out and subsequent drowning.

Breast Stroke

Use this stroke to swim underwater, through oil or debris, and in rough seas. If you are a good swimmer and not wearing combat gear, the breast stroke is the best stroke for long-range swimming because it provides good visibility and allows you to conserve your energy and maintain a reasonable speed.

Body position
Lie prone in the water. Swim with your trunk and legs projecting back and down at an angle of 20 to 30 degrees.

Extend arms out in front (hands together [side by side]), and extend legs behind (toes pointed) to prevent drag.

Face downward, looking forward at a 45 degree angle to break the water and to prevent water from washing into the collar area causing drag.

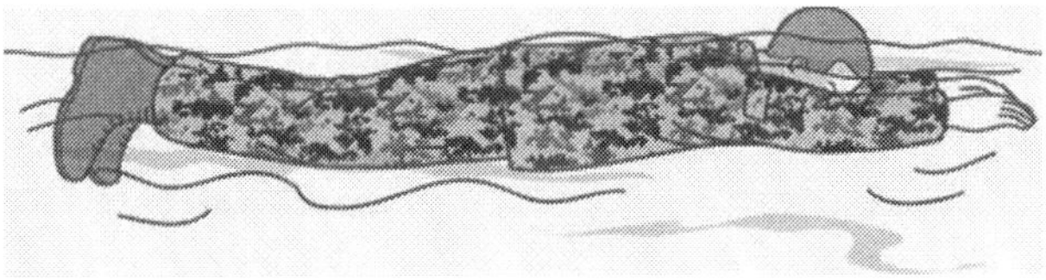

This is known as a glide.

Arm action

Turn your palms outward and bend your arms slightly.

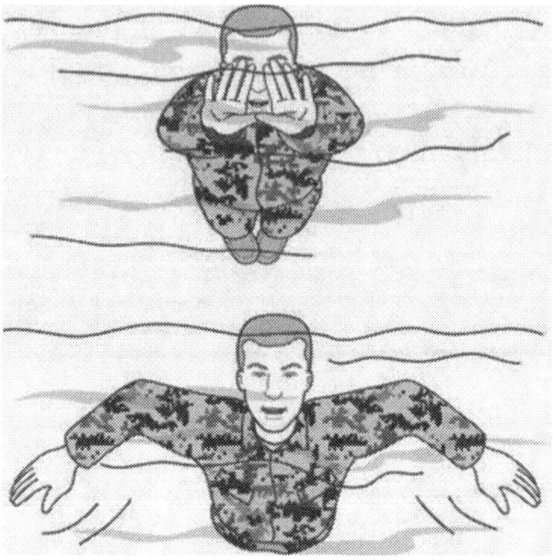

Sweep your arms sideward and slightly downward until your hands are opposite and slightly below your shoulders.

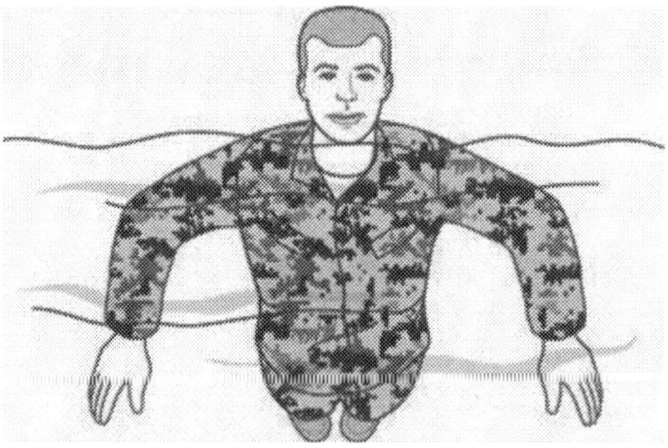

Rotate your head up, breathe once your mouth breaks the surface.

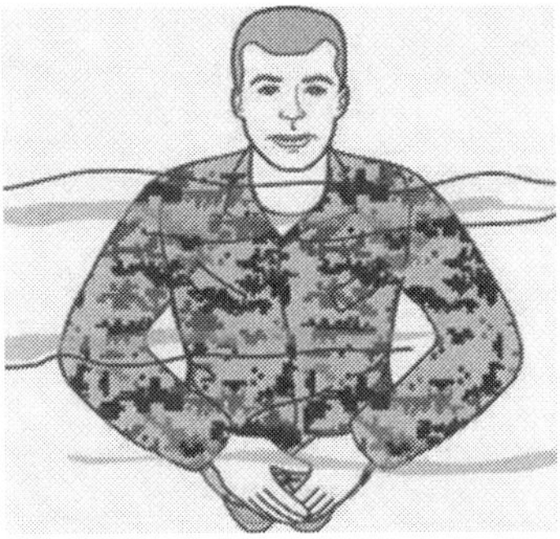

Bring your hands and arms up along your chest and thrust them forward until they are extended and ready to execute the next arm pull.

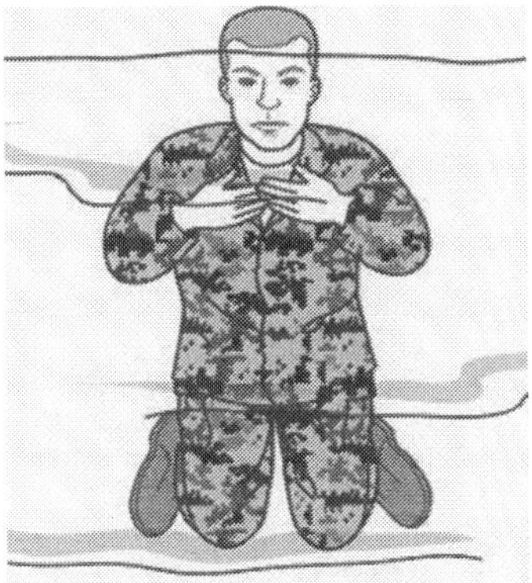

As the arms start their recovery into the glide, the head should rotate forward, resubmerging the face.

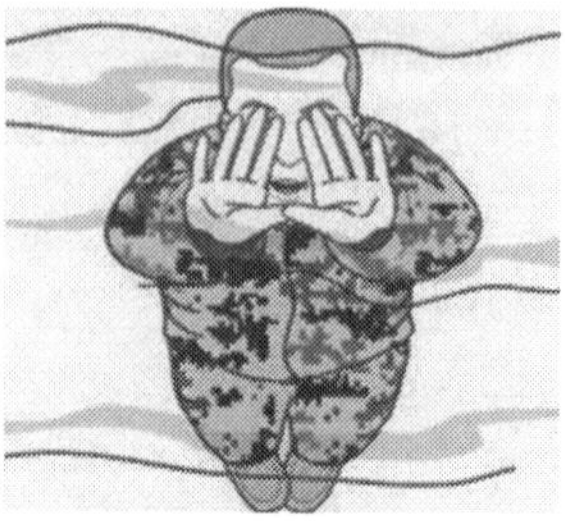

Leg action

Draw your heels toward your buttocks, establish a 45 degree bend in the knees.

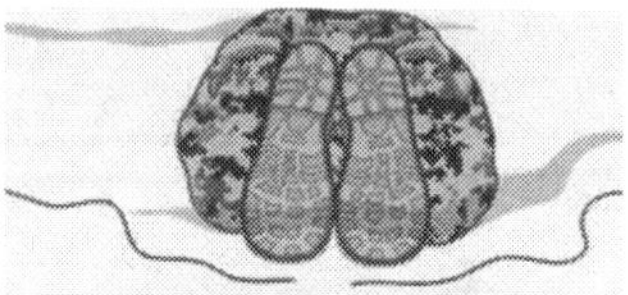

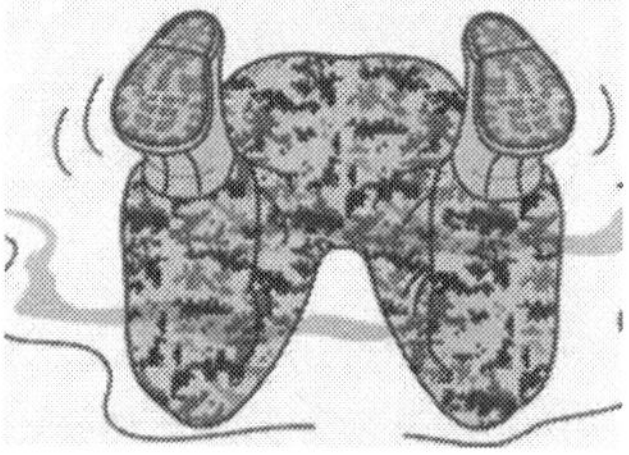

Thrust your legs outward and rearward, then squeeze them together. The whipping action of the feet aids forward propulsion.

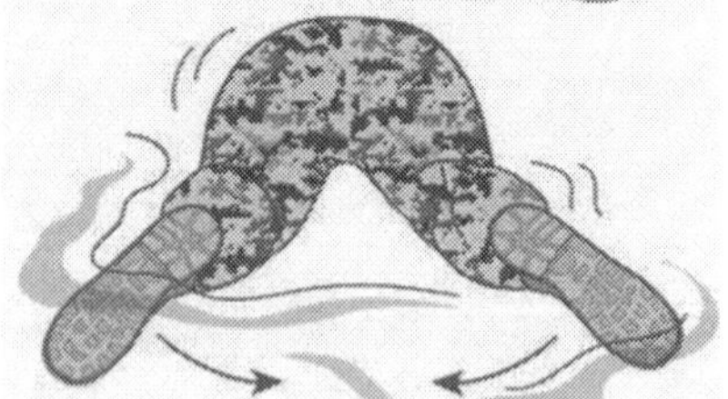

This is known as the breast stroke kick.

Breathing
Inhale during the arm pull and exhale through your mouth and nose during the finish of the breast stroke kick and glide.

Coordination
The stroke movement is in three counts:

Begin your arm pull. Near the finish of the pull, flex your knees and bring your heels toward your buttocks. The arm pull counteracts the resistance created by the knees.

As the arm pull is completed, thrust your hands forward, kick your legs outward and rearward, and squeeze them together.

Glide through the water for approximately 1 to 3 seconds or until your forward momentum decreases, then begin the next stroke.

Side Stroke

The side stroke is a survival stroke because you use both arms for buoyancy, with each arm creating a slight propulsion. The majority of your body's propulsion comes from your kick. To execute the side stroke—

Body position
Lie on your side with your lead (bottom) arm extended beyond (with a slight bend in your elbow) your head and in line with your body. Palm is down and your hand is submerged 6 to 8 inches.

Extend your trail (top) arm down the length of your body over your thigh.

Keep your legs straight and together, toes pointed rearward.

Keep your face out of the water, this allows for free breathing.

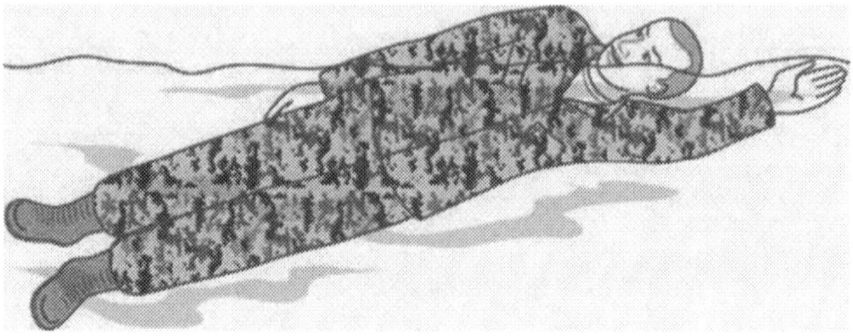

This is known as the glide.

Arm action

With your lead arm, pull your arm downward, while flexing at the elbow, until it is straight down from your shoulder.

Rotate your shoulder and pull your elbow into your side. This should put your lead hand at shoulder level. At the same time, turn your palm toward your face and thrust forward to your original, extended position.

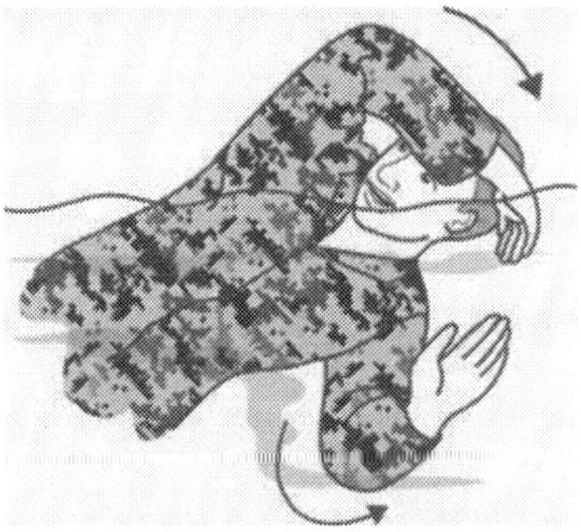

Draw your right hand upward in front of your chest to shoulder level. Rotate your palm toward your feet, then push it downward in front of your body toward your feet to catch the water.

Push your trail hand back ward to its original position on top of your thigh. (Your trail hand starts forward and meets your lead hand at your chest/shoulder.)

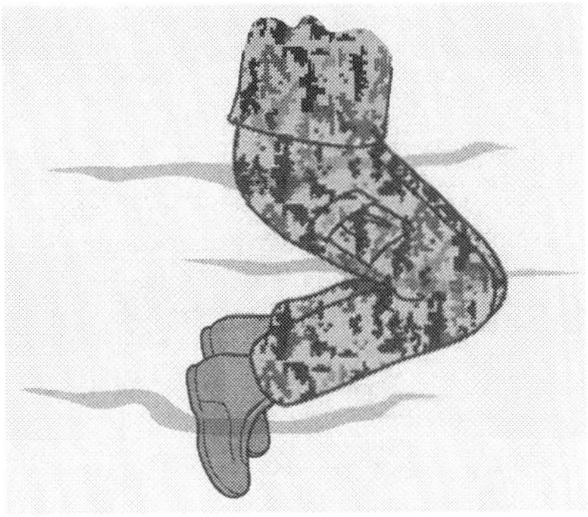

Leg action

To perform the scissor kick, the top leg always goes forward and the bottom leg always goes rearward. From the extended position, draw—or recover—your feet toward your buttocks until your legs are bent at a 45 degree angle at the knees and the hips are flexed at a 45 degree angle with the thighs.

Once the legs have completed their recovery and while maintaining a 45 degree bend in the knees, extend the legs fully into a "V" shape in order to catch the water for the propulsion phase.

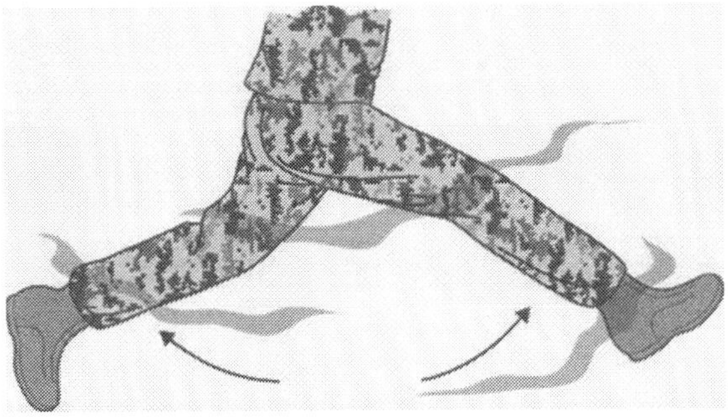

Once the legs are separated and extended forward and rearward to the "V" position, sweep the legs together until the feet are together.

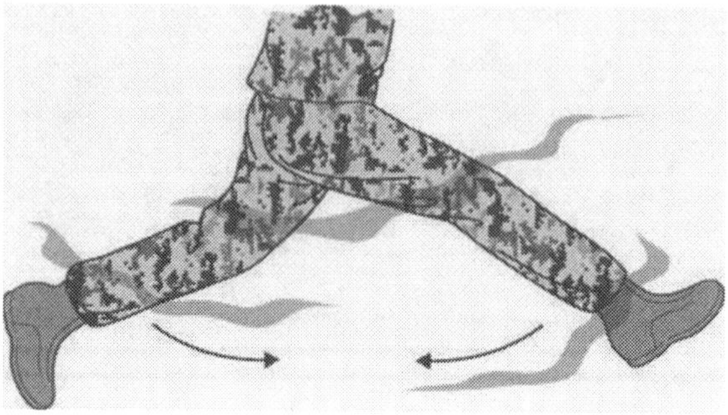

You are now in the glide position.

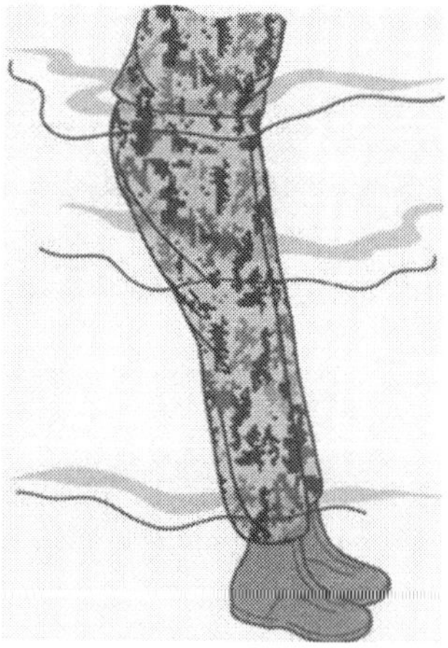

Breathing

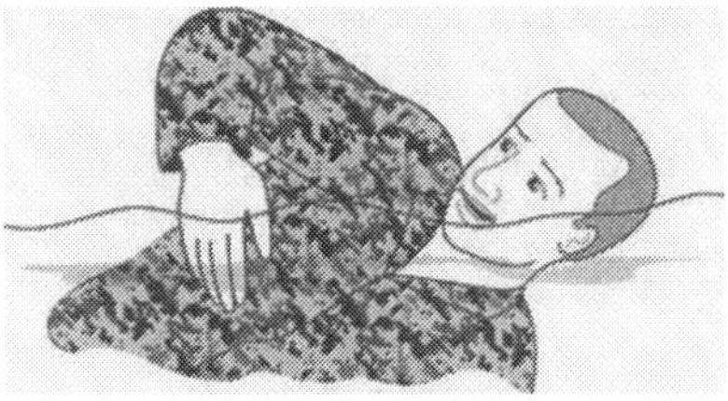

As long as your face remains clear of the water, it is a free breathing stroke. However, it is recommended that you exhale then quickly inhale when the legs are sweeping back together in the scissor kick. This is when the body reaches its highest point in the water, thus clearing the face completely from the surface of the water making it the optimum time to breathe.

Coordination

Begin the stroke with the downward pull of your lead arm. At the same time, bring your trail arm upward and draw your knees up to begin the kick. Let the thrust of the lead arm, push of your trail arm, and the kick of your legs coincide in order to finish the glide position. Glide through the water for approximately 1 to 3 seconds or until your forward momentum decreases, then begin the next stroke.

Elementary Backstroke

The elementary backstroke is also an excellent survival stroke. It relieves the muscles that you use for other strokes, and it is the recommended stroke for weak swimmers or nonswimmers. To execute the elementary backstroke—

Body position
Start on your back.

Face up, chest up, and hips up, keeping an arch in your lower back with arms pressed to your sides and your legs extended and joined to prevent drag.

Arm action
Trace your hands up your sides to an area near your armpits then extend your arms out to the sides to form the letter "T" (palms facing feet), locking out the elbows.

> NOTE: Don't raise your arms above your head. This creates drag, changes your body position, and submerges the head.

Slap your palms to your thighs using a strong sweeping motion.

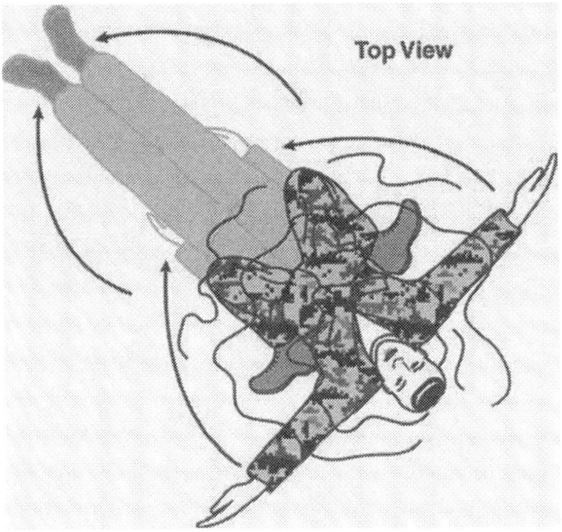

Leg action
Bend both legs at the knee (90 degree angle) slightly separating your knees and drawing your heels downward to a point under and outside your knees. The knees are spread as wide as the hips or slightly wider depending on the body type of the swimmer.

Circle around in a whipping action, ending with legs in a glide position.

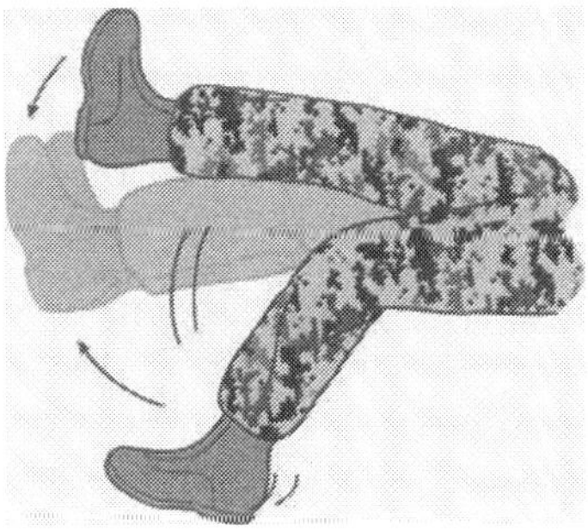

Breathing

Breathe anytime during this stroke. However, it is recommended that you exhale then quickly inhale when your arms are sweeping back toward your sides and while your legs are sweeping back together. This is when the body reaches its highest point in the water, thus clearing the face completely from the surface of the water and making it the optimum time to breathe.

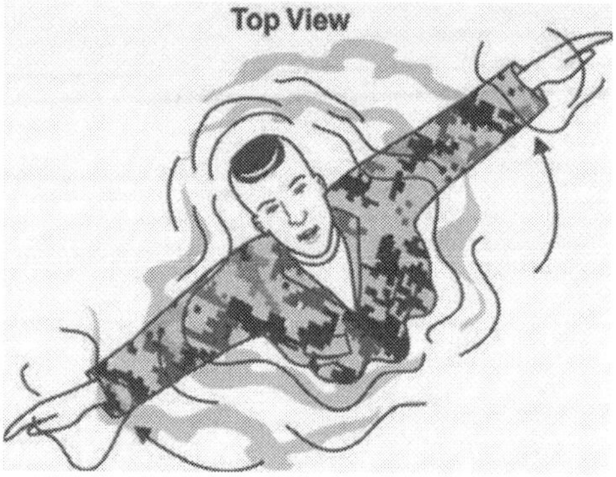

Coordination

The stroke movement occurs in three counts (recovery, catch, power).

Begin the arm pull (recovery).

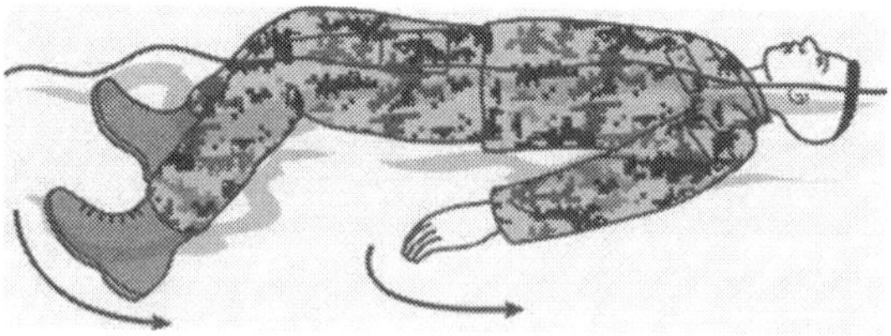

Near the finish of the pull, flex your knees to a 90 degree angle. The arm pull counteracts the resistance created by the knees.

Kick out your legs, and squeeze them together as the arm pull is completed (catch, power).

Glide through the water for 1 to 3 seconds or until your forward momentum decreases, then begin the next stroke as your momentum slows.

Chapter 2
Water Rescues

A drowning victim can panic and react with unexpected violence and can seize and inadvertently drown a rescuer. Therefore, if possible, a water rescue should be executed from a distance. Reaching, wading, or throwing methods are used in a distance rescue.

If a victim is too far away to use these methods, a swimming rescue may be necessary. Remove all combat gear, when possible, before entering the water. Swim within 2 to 6 yards of the victim to maintain a margin of safety, this allows you to reassess the situation and reassure the victim. If the victim is unconscious, use the wrist tow method or cross-chest carry method to pull the person to safety. If the victim is struggling, use a rear approach and then execute either a single armpit level off or a double armpit level off before towing the victim to safety.

If the victim does begin to overpower you, there are techniques that allow you to defend yourself without having to abandon the rescue. These techniques include the block, the wrist-grip escape, the front head-hold escape, and the rear head-hold escape. These techniques allow you to separate yourself from the victim, reassess the situation, and then attempt the rescue again.

> *Note:* In this chapter's illustrations, the rescuer is shown without a helmet.

Reaching Rescue Techniques

Reach

From a safe position at the water's edge, reach out to the victim. Talk constantly to calm the victim. Retain partial contact with land or some solid support structure (e.g., pier, bridge). If the victim is close but still beyond reach, extend an object (e.g., stick; pack; rifle with magazine removed, chamber empty, and muzzle pointing toward victim) that the victim can grasp. Pull the victim slowly to safety. Once the victim is close to shore, remove the victim from the water.

NOTE: You can also extend a foot to the victim if you can retain a secure grip on a solid support structure.

Reach From a Deck

Reach from a deck rescue can be executed by either a swimmer or a nonswimmer, and it can be used on an active or a passive victim. Be sure to reassure the victim during the rescue. To execute reach from a deck—

Lie prone on the deck with your body firmly anchored. To anchor your body, lie flat, spread legs apart, and extend one arm behind you with your palm down and on the deck.

Keep as much of your weight on the deck as possible and extend your free hand to the victim.

Grasp the victim's wrist from above, your thumb and index finger are facing you. The victim should never be allowed to grab you and put your life at risk. Therefore, never reach across the victim to grab his wrist, always grab the wrist that is the closest to you.

Keep your arm straight and locked out and pull the victim to the side of the deck. The victim should never be allowed to grab you and put your life at risk. Therefore, never pull the victim into you, always pull the victim into the side.

Arm Extension

An arm extension rescue can be used if you cannot reach the victim using the reach from a deck rescue technique and you must enter the water. An arm extension rescue technique is used for a victim who is either active or passive. Once you determine that a reach from a deck rescue technique is not viable—

Reassure the victim and quickly ease into the water while holding onto the deck with one hand.

Grasp the victim's wrist from above, your thumb and index finger are facing you. The victim should never be allowed to grab you and put your life at risk. Therefore, never reach across the victim to grab his wrist, always grab the wrist that is the closest to you.

Keep your arm straight and locked out and pull the victim to your side. The victim should never be allowed to grab you and put your life at risk. Therefore, never pull the victim into you, always pull the victim into the side.

Leg Extension

If the victim is beyond the reach of your arm, ensure that you have a firm grip on the deck, extend a leg to the victim and allow him to grab it, slowly bring the victim in closer until you can grab the victim's wrist that is holding onto your leg. Once you have a firm grasp on the victim's wrist, use the steps in the arm extension to pull the victim to safety. The leg extension rescue can only be used to rescue an active victim because the victim must be able to grab the rescuer's leg.

Wading Assist

Do not wade into water that is deeper than your chest. Talk constantly to calm the victim. If possible, do not touch the victim directly. Extend an object (e.g., stick; pack; rifle with magazine removed, chamber empty, and muzzle pointing toward victim) that the victim can grasp. Once the victim grasps the object, pull the victim slowly to safety.

Throw

If the victim is not within reach, use an expedient line to throw a lifesaving device to the victim. A lifesaving device can be any weighted item that floats (e.g., a canteen that is one fourth full of water). The lifesaving device must be secured to the end of the rope so that the rope will feed out from its coil when tossed. Talk constantly to calm the victim. Once the victim grasps the line or the lifesaving device, pull the victim toward you at a steady pace that keeps the victim's head above the water's surface. DO NOT pull so strongly that you break the victim's grip on the line. To prepare and use an expedient line and lifesaving device—

Tie a bowline at one end of the rope. (The appendix illustrates various types of knots.)

Unfasten the lid of a canteen.

Place the bowline around the neck of the canteen.

Refasten the lid so the canteen hangs from the bowline loop.

Place one end of the rope under the ball of your forward foot to secure it, either tie a knot at the end of the rope or tie an object at the end of the rope to create a block. Stand with your weight on the end of the rope. DO NOT tie the rope around your ankle.

Coil 20 to 30 yards of the rope, and hold it in a nonthrowing hand.

Place the canteen in your throwing hand.

Use an underhand throw to pitch the canteen and rope a short distance over the victim's head. Keep your nonthrowing hand open so the coil can unfold freely. The rope should trail across the victim's outstretched arms.

Retrieve the rope if the throw is inaccurate or the victim fails to grasp it. Recoil the rope as it is retrieved.

Divide the coil and throw again.

Lifesaving Approaches

Properly approaching the victim is as important as any other aspect of a rescue. Knowing the appropriate and safest method needed to approach a victim is critical in ensuring your safety and the victim's survival. Determining the victim's physical state (distressed swimmer, victim that is active and drowning, or victim that is passive and drowning) is crucial and will determine which type of approach you will execute.

Front Surface Approach

The front surface approach is typically performed when the victim is passive. You must remember that approaching the victim from the front is dangerous because a distressed swimmer or victim that is active can lunge toward you.

Rear Approach

If the victim is active, you must remember that his response can change rapidly. Therefore, approaching him from the front could be extremely dangerous. The victim could easily grab you and take you under the water during a state of panic. Therefore, use a rear approach when possible. When approaching a victim from the rear, reassure the victim until you get to within 2 to 6 yards of the victim; at that point, stop talking to the victim in order to maintain an element of surprise.

Approach Strokes

Before approaching a victim, you must evaluate the victim:

- Is the victim a distressed swimmer?
- Is the victim active and drowning?
- Is the victim passive and drowning?

Select your approach stroke based on your evaluation. Remember, the safety of the rescuer is paramount. Do not endanger yourself in an attempt to reach a drowning victim.

Crawl Stroke Approach Stroke

The crawl stroke (free style) is the fastest approach stroke. This stroke is used when the victim is passive and/or unconscious in the water. That is, the victim is face down, submerged or near the surface of the water, is not breathing, or is not moving. If any of these conditions exist, there is an obvious need to reach the victim as quickly as possible.

Execution of the crawl stroke approach stroke is the same as the crawl stroke (see pages 1-40 through 1-44), except that your head remains above the water's surface to allow for free breathing and maintaining eye contact with the victim.

Maintaining eye contact is critical. If the victim becomes submerged, you will have a better chance of locating him if you know his last location. Your distance from the victim will determine if your head remains out of the water the entire time you are executing the stroke. If you are 54 yards or less from the victim, you should maintain eye contact by keeping your head raised until you reach the victim. If you must swim more than 54 yards to reach the victim, make eye contact with the victim, place your face back in the water for four or five strokes, then raise your head again to regain eye contact with the victim (repeat these steps until you reach the victim).

Body position

The position of your body is horizontal to diagonal and influenced by the position of the head. This allows you to maintain eye contact with the victim.

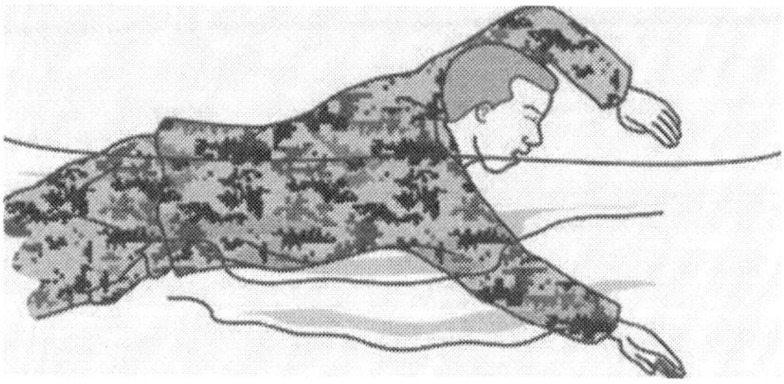

Arm action

While one arm is forward of your head to catch the water for the propulsion phase, extend your other arm alongside your body in what is known as the recovery phase. The recovery phase ends when your arm is extended forward of your head to begin the propulsion phase. In the propulsion phase, your arms are approximately shoulder-width apart and the pulling action is slightly wider and deeper to compensate for your raised head.

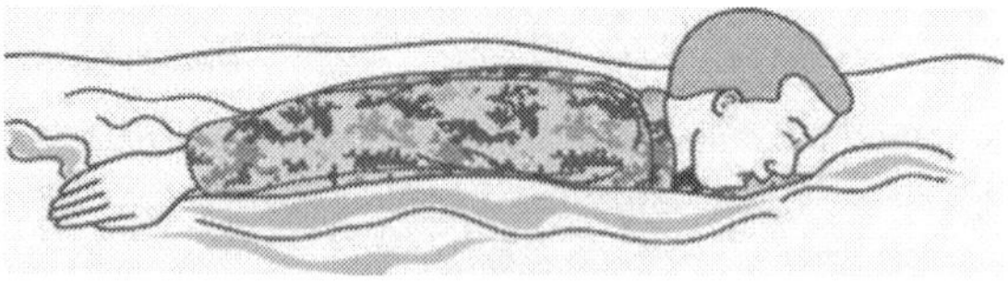

Leg action
Kick one leg in a downward motion while your other leg is recovering to the surface to prepare for the next kick (this is known as a flutter kick). This kick is used for propulsion and maintaining the lower body as horizontal to the water's surface as possible. When your head is raised, your knees are slightly bent in order to keep your feet near the surface.

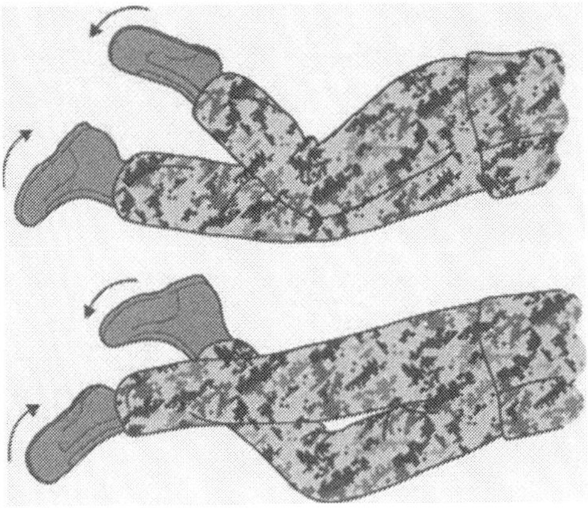

Breathing
If your face is above the water, breathe as needed (free breathing).

If traveling in excess of 54 yards to reach the victim, rotate your head to the right or left during the recovery phase and inhale when your face is out of the water or wait until your head is out of the water and facing forward while regaining sight of the victim.

Coordination
This stroke uses constant arm and leg action.

Breast Stroke Approach Stroke

The breast stroke approach stroke is used when a swimmer is distressed or a victim is active and drowning. This stroke is also used when the victim has a suspected spinal injury because it minimizes the wake created around the victim, minimizes the movement of the victim's head and neck, and helps prevent further injury.

If a victim is at a distance greater than 54 yards and if the swimmer is in distress or the victim is active and drowning in open water (ocean, lake), you should use the crawl stroke approach stroke to rapidly approach the victim, stopping approximately 11 yards from the victim in order to assess the situation.

Execution of the breast stroke approach stroke is the same as the breast stroke with a few minor deviations.

Body position
The position of your body is horizontal to diagonal and influenced by the position of the head. This allows you to maintain eye contact with the victim. It also allows you to communicate with the victim in order reassure him as you approach.

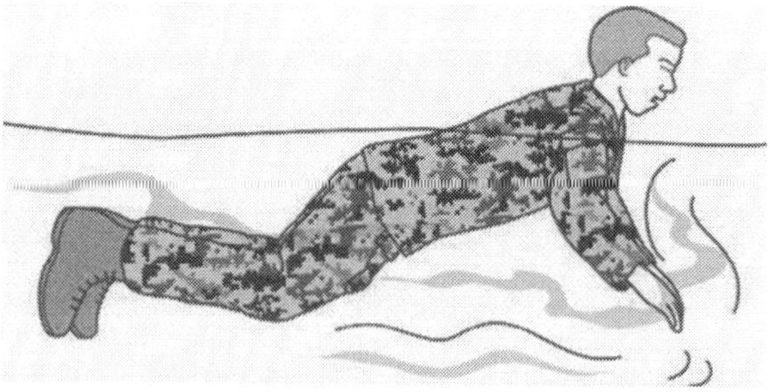

Arm action

Execute the arm action for the breast stroke, except that your arm pull is wider and deeper, which allows your head to remain above the water's surface.

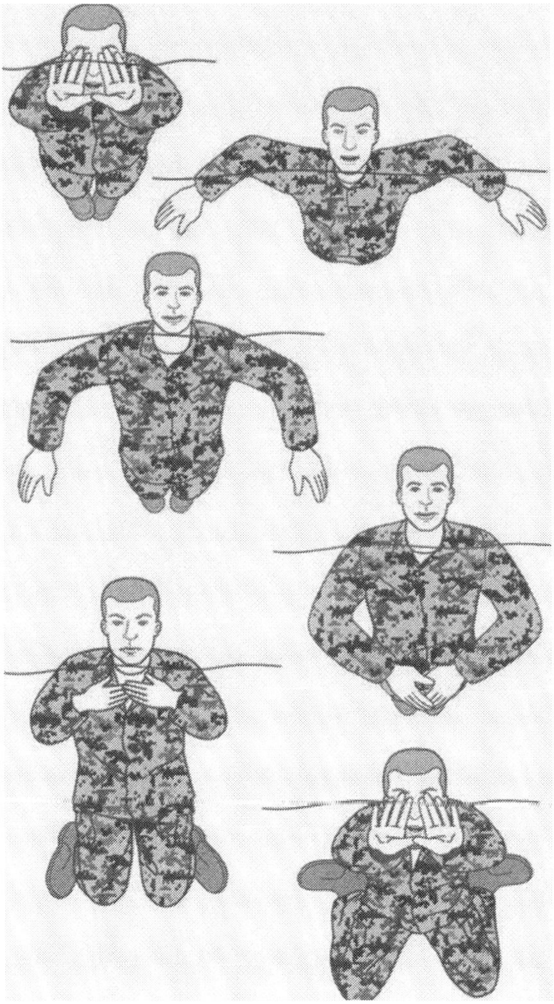

Leg action
Execute the leg action for the breast stroke, except that your legs open wider during the propulsion phase of the stroke.

Breathing
Keep your face above the water and breathe as needed (free breathing).

Coordination
Coordination is the same as the breast stroke, except that there is no glide period. You constantly stroke until you reach the victim.

Level Offs

A properly performed level off positions the victim's face up and horizontal to the water's surface. There are three types of level offs: front surface approach, single armpit level off, and the double armpit level off. The single armpit level off and the double armpit level off are done by performing a rear approach, which is the preferred approach when a victim is active.

Front Surface Approach

The front surface approach is performed when the victim is passive. Approaching a victim from the front may place you in danger because a distressed swimmer or a victim who is active may lunge toward you. To execute a front surface approach—

Stop 2 to 4 yards from the victim to reassess the situation and reassure the victim.

Determine which arm to use in order to rotate the victim into a face-up position (your right hand on the victim's right wrist or your left hand on the victim's left wrist).

Move into position and prepare to grasp the victim.

Turn sideways and move toward the front of victim.

Once in position, reach forward and grab the victim's wrist. Your thumb is on the inside of the victim's wrist, as if you were checking his pulse with your thumb. Your remaining fingers wrap around the victim's wrist.

Lean back immediately and execute a powerful scissor kick or inverted scissor kick and perform short, vigorous pulls with your free arm.

As the victim begins to move forward in the water, kick, pull, and twist outboard on the victim's wrist. The momentum created from the kick and the pulling and twisting action of the victim's wrist will rotate the victim into a face-up and horizontal position in the water.

Extend and lock out your towing arm down the length of your body and execute a wrist tow to move the victim to safety.

Single Armpit Level Off

Approach the victim.

Stop 2 to 6 yards from the victim to reassess the situation, reassure the victim, and maintain your margin of safety.

Approach the victim slowly and grab his armpit with your hand (your right hand to victim's right armpit or your left hand to victim's left armpit). Position yourself sideways to the victim and place your elbow in the center of his back.

Pull with your hand that is in the victim's armpit while pushing with your elbow that is in the victim's back to place the victim horizontal on the water's surface (face up). To assist in placing the victim horizontally, use your free arm to execute short, vigorous pulls and use your legs to execute a scissor kick or inverted scissor kick.

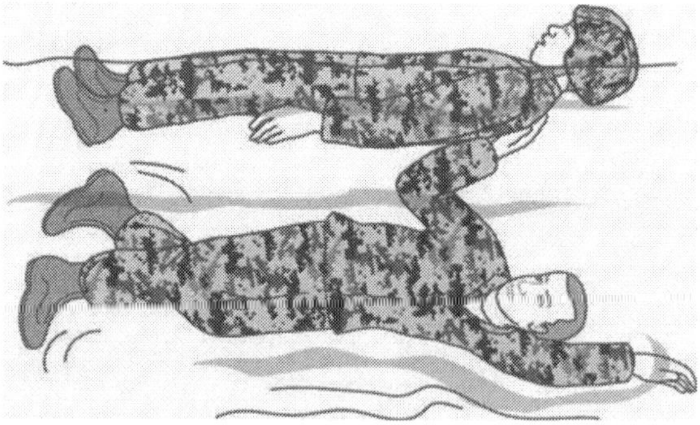

Once the victim is horizontal and on the water's surface, begin forward momentum by extending your towing arm to a fully locked-out position and executing a single armpit tow.

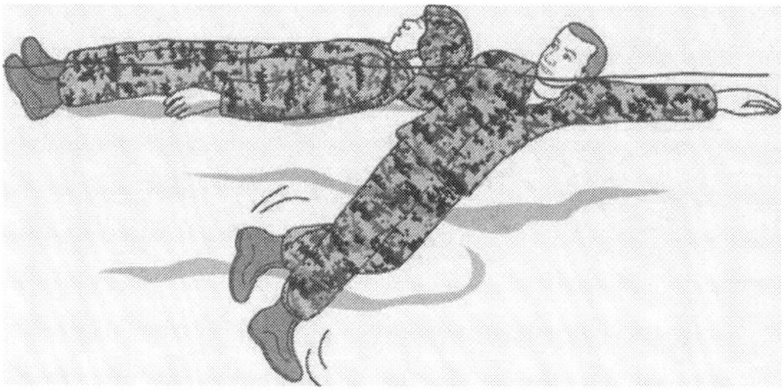

Double Armpit Level Off

Approach the victim.

Stop 2 to 6 yards from the victim to reassess the situation, reassure the victim, and maintain your margin of safety.

Approach the victim slowly and grab his armpits with both your hands. Place your elbows on his back.

Pull with your hands that are in the victim's armpits while pushing with your elbows that are in the victim's back and use an inverted breast stroke kick to place the victim horizontal on the water's surface (face up).

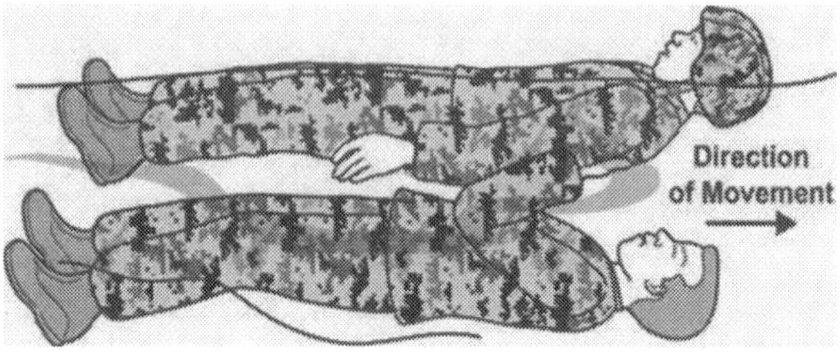

Once the victim is horizontal and on the water's surface, begin rearward momentum by slowly extending your arms into a fully locked-out position and executing a double armpit tow and inverted breast stroke kick.

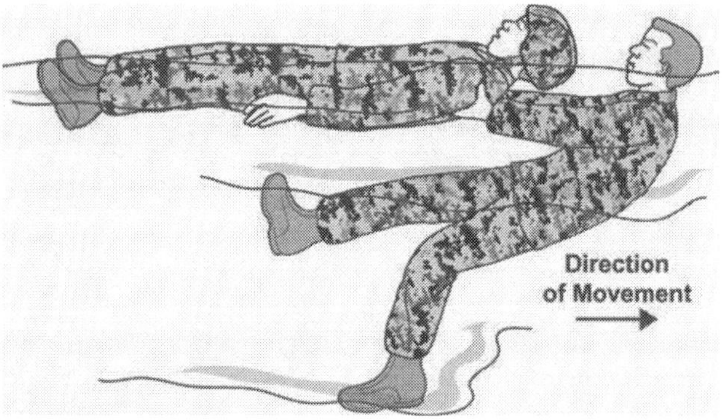

Rescue Techniques

Wrist Tow

Use the wrist tow method to rescue a victim who is floating face down. DO NOT use the wrist tow on a struggling victim. If time allows, remove your helmet and gear before attempting the rescue. Swim toward the victim using a modified breast stroke. Swim within 2 to 6 yards of the victim to maintain a margin of safety, this allows you to reassess the situation and reassure the victim. The following steps show proper front surface approach, wrist tow procedures:

Approach the victim from the front and grasp the underside of the victim's left wrist with your left hand or the right wrist with your right hand. Ensure that your thumb is on the underside of victim's wrist.

Lean back, pulling and kicking strongly to move the victim into a horizontal position.

Twist the victim's wrist to rotate the victim into a face-up position.

Swim toward safety using the lifesaving stroke.

> NOTE: The lifesaving stroke is a modified side stroke wherein the top arm is used to tow or carry a victim to safety and use a scissor kick or an inverted scissor kick is used for propulsion.

Keep a firm grip on the victim's wrist.

Keep your towing arm fully extended and along your side. This keeps the victim in the water column and prevents drag.

Ensure the victim's head does not go under water during the recovery.

Single Armpit Tow

You perform the single armpit tow after a victim has been properly leveled off. The single armpit tow uses the lifesaving stroke. To execute the single armpit tow—

Place one hand (top arm) in the victim's armpit. Your towing arm is straight and along your side.

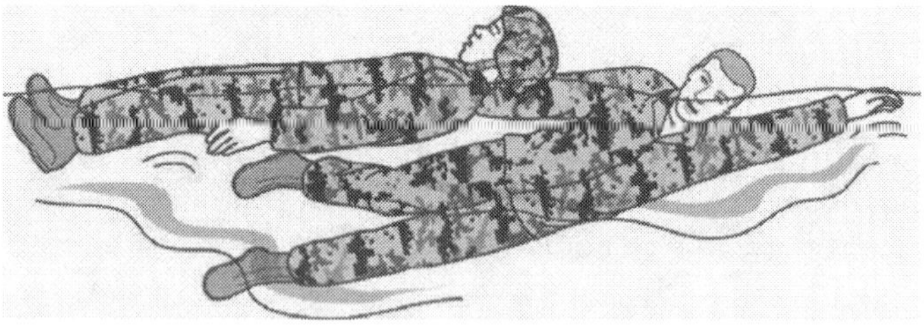

Use your lead arm (bottom arm) to execute short, vigorous pulls.

Execute a scissor or inverted scissor kick in a continuous and vigorous manner.

Use either free breathing or explosive breathing during the tow. During free breathing, keep your head above the water's surface and continuously reassure the victim. During

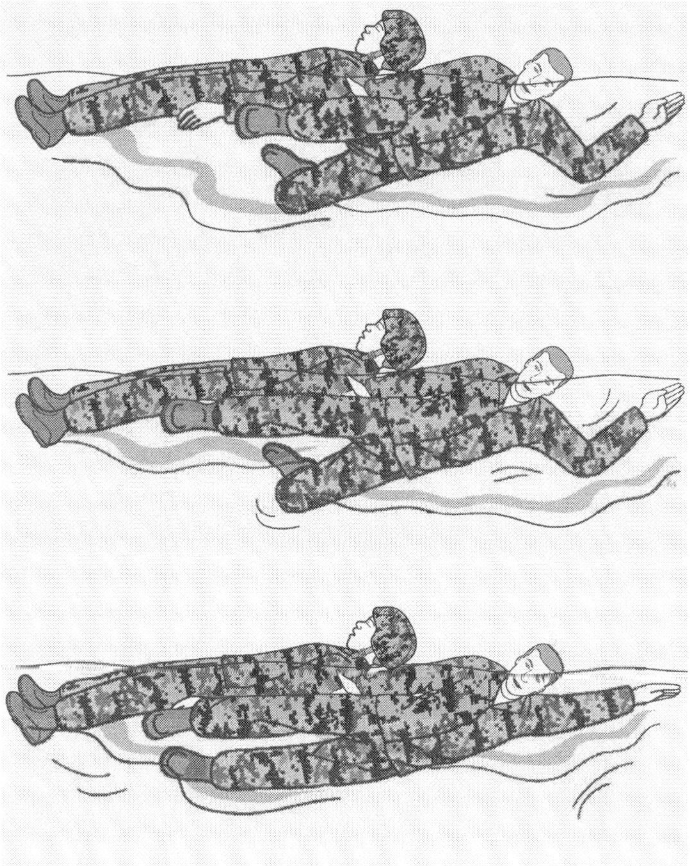

explosive breathing, put your head under the water to plane off your body's angle and reduce drag. Execute approximately two strokes, then lift your head up to breath and reassure the victim. Return your head back underwater for the next few strokes.

Double Armpit Tow

You perform the double armpit tow after a victim has been properly leveled off. To execute the double armpit tow—

Place both hands in the victim's armpits.

Extend both arms fully along your body. You are on your back, and your face is clear of the water.

Use free breathing.

Reassure the victim at all times.

Use the inverted breast stroke kick (it is the only kick that can be used due to your body position) to tow the victim to safety. (The inverted breast stroke kick is the same kick used in the elementary backstroke.) Your kick must be continuous and vigorous in order to keep the victim's face above the water.

Collar Tow

You perform the collar tow after a victim has been properly leveled off. To perform a collar tow using the victim's blouse—

Maintain control of the victim by grasping his armpit with one hand and then with your free hand grasp either his combat gear or his blouse between his shoulder blades. If grasping the blouse, grasp the material with your palm up, then turn your hand over to tighten the material.

Release the victim's armpit once control is established and execute the lifesaving stroke to tow the him to safety.

Cross-Chest Carry

Use the cross-chest carry to carry a victim to safety if the victim is struggling or when moving through heavy surf. Remove your helmet and gear before attempting a rescue. Talk constantly to calm the victim. Swim toward the victim using a breast stroke approach stroke. Swim within 2 to 6 yards of the victim to maintain a margin of safety, this allows you to reassess the situation and reassure the victim. The following steps illustrate proper cross-chest carry procedures:

> **CAUTION:** The cross-chest carry causes fatigue even if you are in excellent physical condition.

Use a level-off technique to place the victim in a horizontal, face-up position.

Retain a grip on the victim with one hand. Reach over the victim with your free hand to encircle the victim's chest. Place your free hand on the victim's opposite rib cage, just below his armpit.

Release your grip once you have a secure hold on the victim's chest.

Swim toward safety using the lifesaving stroke while keeping a firm grip on the victim's chest and your hip on his back.

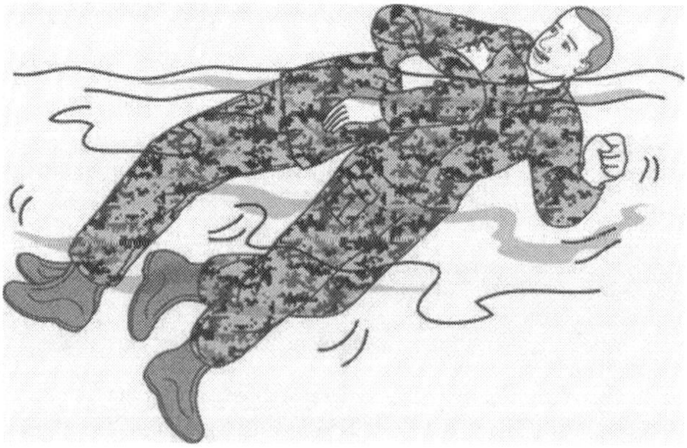

This procedure brings the victim's face and shoulders clear of the water, and typically the victim stops struggling. Sometimes, however, the victim will struggle during the swim to safety. If this happens, either tighten your grip on the victim or defend yourself with one of the techniques discussed on pages 2-25 through 2-30.

Tired Swimmer's Assist

You must maintain a 2 to 6 yards margin of safety from the victim at all times while you are getting into position to perform the tired swimmer's assist. Once behind the victim, you extend one arm, hand straight. Place your straight hand underneath the victim's opposite armpit. Maintain a 45 degree angle away from the victim, your arm is locked out. Assist the victim in propulsion until both of you reach safety.

Defense Against a Drowning Victim

DO NOT sacrifice your life in an attempt to save the victim. A struggling drowning victim poses great danger to anyone nearby. Driven by panic, the victim can grab you with great strength in an effort to climb out of the water. This can result in death for both you and the victim. The following techniques can be used to defend against a drowning victim, but the best defense against attack by the victim is to stay out of reach.

Block

The block prevents the victim from grabbing you if you have approached from the front. If the victim lunges toward you, react as follows:

Place one or both open hands against the victim's upper chest; being careful to avoid the victim's face, neck, and abdomen.

Lean backwards and submerge rapidly. Keep your blocking arm(s) extended.

Swim underwater and away from the victim, and return quickly to the surface.

Stop 2 to 6 yards from the victim to reassess the situation.

Determine an appropriate course of action.

Wrist-Grip Escape/Wrist-Grip Escape Alternative

The wrist-grip escape is used when a victim grabs your arm or wrist. Submerge the victim quickly by reaching across with your free hand, pushing down on the victim's shoulder to submerge him, and kicking to propel yourself upward. While keeping the victim submerged with your hand on his shoulder, give three hard jerking pulls with your trapped hand in an attempt to break free from his grasp. Once free, swim clear of the victim and reassess the victim's condition.

If the wrist-grip escape is unsuccessful in freeing your hand, use the wrist-grip escape alternative: use your free hand to grab your trapped fist, rotate thumbs up, apply bone-on-bone contact with the victim's arm, pry your hand out of his grasp, and quickly swim away from the victim.

Front Head-Hold Escape

The front head-hold escape allows you to escape when you are facing a victim who is gripping you around your head and neck. To execute the front-head hold escape—

Take a quick breath and tuck your chin into your shoulder to protect your throat.

Clap your hands above your head (three times) to submerge instantly. This drags the victim below the water, lifts his arms from around your neck, and, typically, he releases his grasp in order to get back to the surface. If he doesn't release his grasp, apply pressure to the victim's brachial pressure points (which are located inside of the upper arm, above the elbow).

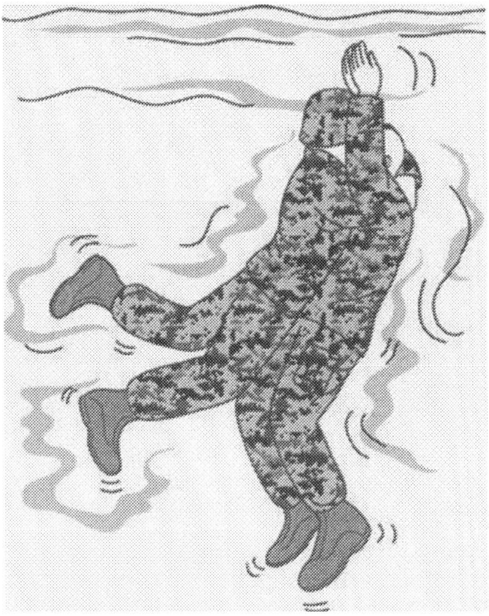

Thrust the victim's arms up and away.

Keep your chin tucked to protect your throat, and swim underwater away from the victim and return quickly to the surface at the ready position.

Stop 2 to 6 yards from the victim to reassess the situation.

Determine an appropriate course of action.

Rear Head-Hold Escape

The rear head-hold escape allows you to escape when a victim is gripping your head and neck from the rear. To execute the rear head-hold escape—

Take a quick breath. and tuck your chin down, turn your head to either side, and raise your shoulders to protect your throat.

Take a strong stroke, clap your hands above your head (three times), and submerge instantly. This drags the victim below the water and, typically, he releases his grasp in order to get back to the surface. If he doesn't release his grasp, apply pressure to the victim's brachial pressure points (which are located inside of the upper arm, above the elbow).

Thrust the victim's arms up and away.

Twist your head and shoulders until free.

Swim underwater away from the victim and return quickly to the surface.

Stop 2 to 6 yards from the victim to assess the situation.

Determine an appropriate course of action.

Administering First Aid/Rescue Breathing

To administer first aid/rescue breathing, remove the unconscious victim from the water when possible and stop any bleeding.

First, open the airway and check for breathing. If the victim is not breathing, give two breaths of air. Check for a pulse. If there is a pulse but the victim is not breathing, continue rescue breathing. If there is no pulse, start cardiopulmonary resuscitation (CPR). For detailed information on administering CPR and rescue breathing, refer to American Red Cross CPR skill cards.

WARNING

If the victim has no pulse and is not breathing, administer CPR immediately. If the victim does have a pulse but is not breathing, give rescue breathing only. If the victim has a pulse and is breathing, DO NOT give CPR—CPR could prove fatal.

Next, protect any wounds from exposure, and treat for shock.

Chapter 3
Treatment of Casualties and Avoidance of Dangerous Marine Life

To survive in the water, you will face many challenges other than just being able to swim. For example, injuries and/or fatigue can lead to drowning and exposure to the elements can lead to hypothermia or heat injuries and may require medical attention. Finally, your presence in the water may be seen as a threat to or food source for a variety of marine life and you may be attacked. This chapter advises you on how to minimize these risks in order to stay alive.

Drowning

Drowning is suffocation by a liquid. A drowning victim inhales water into the lungs or the throat closes by reflex so that little or no water can enter the windpipe. In either case, a victim can no longer breathe.

Symptoms

One symptom of drowning is that the victim may call for help and has an expression of dread or panic. But typically a victim that is active and drowning may not call for help because he is trying to conserve his air and will not speak.

Another symptom of drowning is when the victim thrashes at the water's surface. If the victim stops or grows calmer, he has likely been overcome by fatigue, hypothermia, or a lack of air. At this stage, the victim usually has 20 to 60 seconds before going under the water's surface.

Treatment

If the victim is not breathing, begin rescue breathing. Place the victim on his back, tilt head back to open airway, pinch the nose, and give two full breaths. If the victim does not inhale during the first two breaths, reposition his head and attempt two more breaths. Check for a pulse. If a pulse is present, but the victim is still not breathing, continue rescue breathing. If a pulse is not present, begin CPR. See MCRP 3-02G, *First Aid*, for rescue breathing and CPR details.

WARNING

> If the victim has no pulse and is not breathing, administer CPR immediately. If the victim does have a pulse but is not breathing, give rescue breathing only. If the victim has a pulse and is breathing, DO NOT give CPR—CPR could prove fatal. Continue first aid until medical help arrives.

A victim who is not breathing and has no pulse may appear dead. DO NOT decide that death has occurred. Continue with the prescribed treatment. A corpsman or medical officer should decide whether the victim can or cannot be revived.

Hypothermia

Hypothermia is the abnormal lowering of the body's internal (or core) temperature to 95 degrees or below. It occurs when the body loses heat faster than it produces heat. A hypothermia victim loses the ability to move quickly, becomes mentally sluggish, slips into semiconsciousness, lapses into a coma, and dies when internal body temperatures drop too low.

The chilling effects of cold air, wind, or water can produce hypothermia. Water poses the greatest threat because it transfers heat 25 times faster than air. Depending on the water's temperature, a victim can succumb to hypothermia within a few minutes. The body's sudden contact with cold water can also set off a body reaction known as the mammalian diving reflex. This reflex can greatly increase survival time (especially for women and children) in or under cold water. The mammalian diving reflex shuts off blood circulation, except for the flow between the heart, lungs, and brain. The small amount of oxygen left in the blood and lungs is saved for the body's vital organs. This reflex has allowed people to survive being under cold water for an extended period of time. Therefore, a cold water drowning victim should be treated as if still alive even though the victim is not breathing, has no pulse, and may appear dead. DO NOT decide that death has occurred. Continue with the prescribed treatment. Victims of hypothermia can appear to be dead when they are not. A corpsman or medical officer should decide whether the victim can or cannot be revived.

WARNING

If the victim has no pulse and is not breathing, administer CPR immediately. If the victim does have a pulse but is not breathing, give rescue breathing only. If the victim has a pulse and is breathing, DO NOT give CPR—CPR could prove fatal. Continue first aid until medical help arrives. Check for a pulse for at least 45 seconds.

Symptoms

Once the body's core temperature drops, the victim will show one or more of the following symptoms:

- Violent and uncontrollable shivering as the body tries to warm itself.
- Slow or slurred speech.
- Disorientation or poor coordination.
- Loss of skin color.
- Blue and pinched lips.
- A slowing or stopping of shivering that progresses into a rigid torso and limbs.

Survival Time

A hypothermia victim's survival depends on the water's temperature and the time spent in the water. A small body build cools faster than a large build. Children cool faster than adults.

To increase your chance of survival in the water, utilize the HELP position described on page 1-37. Extra clothing and inactivity (remaining motionless in the water) can also increase your survival time.

Treatment

A hypothermia victim must be warmed to prevent further heat loss; therefore, treatment should begin as soon as possible. Consciousness of the victim determines the treatment that should be pursued. See MCRP 3-02G for specific treatment information. The following treatment procedures are recommended:

- If the victim is conscious, give the victim warm fluids. Give candy or sweetened foods to a victim who is able to eat.
- If the victim is unconscious, place him on his back with his head tilted back to ensure an open airway.
- DO NOT massage the victim. Massage can break blood vessels and create swelling, internal pressure, and blocked blood circulation.
- DO NOT give alcohol to the victim. Alcohol lowers the victim's body temperature.
- Shock is a possibility, treat accordingly.
- Seek medical help immediately.

If you are able to remove the victim from the water, apply the following steps when possible:

- Get the victim into shelter.
- Remove the victim's wet clothing.
- Put the victim in dry clothing.
- Place the victim in a sleeping bag if one is available. It may be necessary to place another Marine in the sleeping bag with the victim.
- Place as much insulation as possible between the victim and the ground.
- Use hot water bottles, electric blankets, or blankets heated in an oven or by a campfire to warm the victim's neck, groin, and the sides of the chest.

CAUTION: DO NOT apply heat to extremities.

Heat-Related Injuries

Heat-related injuries include heat cramps, heat exhaustion, and/or heat stroke. They can occur when a Marine—

- Is exposed to extreme heat, such as from the sun or a combination of high air temperatures and water temperatures.
- Does not wear proper clothing or gets overexposed to the sun.
- Becomes dehydrated.

With heat cramps, muscles may cramp in the arms, legs, and/or stomach. The victim may sweat excessively. To treat, create improvised shade for the victim and have him drink water.

With heat exhaustion, the victim sweats heavily; presents pale, moist, cool skin; and complains of a headache, weakness, dizziness, and/or loss of appetite. A victim may also experience heat cramps, nausea, vomiting, an urge to defecate, chills, rapid breathing, confusion, and a tingling sensation in the hands or feet. To treat, pour water over the victim and fan him to speed up the coolant effect of evaporation, have the victim drink freshwater, and attempt to provide shade.

With heat stroke, the victim stops sweating, which results in red, flushed, hot, dry skin. The victim may first experience headaches, dizziness, nausea, fast pulse and respiration, and/or seizures and mental confusion. The victim can become unconscious and die if not treated quickly. To treat, create improvised shade for the victim. Have him drink water and elevate both legs if possible.

Burns

Apply a field dressing (or the cleanest material available) to the burn. Give sips of water to a casualty who is conscious and not nauseated. When treating burn casualties—

- DO NOT remove clothing stuck to the burns.
- DO NOT break any blisters.
- DO NOT apply grease or ointments to the burns.

For electrical burns, check for both an entry and exit burn from the passage of electricity through the body. An exit burn may appear on any area of the body, not necessarily opposite the entry burn.

For burns caused by wet or dry chemicals, flush the burns with large amounts of water and cover with a dry dressing.

For burns caused by white phosphorus, flush the area with water, then cover with a wet material, dressing, or mud to exclude air and keep the white phosphorus particles from burning.

For laser burns, apply a field dressing.

Common Medical Problems Associated with Sea Survival

Seasickness

Seasickness is the nausea and vomiting caused by the bobbing motion created by the wave action of a flotation device. Seasickness can result in—

- Dehydration and exhaustion.
- A loss of the will to survive.
- Others becoming seasick.
- Unclean conditions.

To treat seasickness—

- Wash both the Marine and the flotation device to remove the sight and odor of vomit.
- Keep the Marine from eating food until the nausea is gone.
- Have the Marine lie down and rest.
- Give the Marine seasickness pills if available. If the Marine is unable to take them orally, the pills should be inserted rectally for absorption by the body.

Saltwater Sores

Saltwater sores occur when skin that has abrasions or is cut is exposed to saltwater. The sores may form scabs and pus. Do not open or drain. Flush the sores with freshwater, if available, and allow to dry. Apply antiseptic, if available.

Blindness/Headache

Irritants or the effects of the sun's rays reflecting off the water can cause temporary blindness or headaches. If flames, smoke, or other irritants get in your eyes, flush the eyes immediately with saltwater, then with freshwater, if available. Apply an ointment, if available. Bandage both eyes for 18 to 24 hours or longer if the damage is severe. If glare from the sky and water causes your eyes to become bloodshot and inflamed, bandage the eyes lightly. Try to prevent this problem by wearing sunglasses or goggles with a sunglass insert.

Constipation

This condition is a common problem associated with dehydration. For constipation, do not take a laxative if it is available; this causes further dehydration. Drink freshwater, if available.

Sunburn and Dehydration

The sun's rays reflect at all angles off the waves of the water; therefore, sunburn and dehydration are serious problems in sea survival. Try to prevent sunburn by—

- Erecting an improvised canopy, with available floating materials, to provide shade.
- Wearing your soft cover or using a cloth, such as a handkerchief, to cover your head.
- Covering your skin with sunscreen or lip balm from your first aid kit. Your lips, nostrils, eyelids, the backs of your ears, and the skin under your chin sunburns easily. If enough sunscreen cream is available, all exposed skin should be covered.

Dehydration is caused by the loss of the body's vital fluids. Dehydration in saltwater may result from a combination of factors such as a lack of water, the effects of saltwater on skin tissue, sunburn or vomiting from seasickness, and other causes. Sleep and rest and reduced water and food intake are the best ways of enduring periods of exposure. The following measures will delay the effects of dehydration:

- DO NOT drink saltwater.
- DO NOT drink urine.
- DO NOT drink alcohol.
- DO NOT smoke.
- DO NOT EAT unless water is available.

Dangerous Marine Life

You may see many types of marine life around you, and some are more dangerous than others. Generally, sharks are your greatest danger, followed by barracudas. However, most marine life will not deliberately attack a human. The most common injuries from marine life are wounds from bites, stings, or punctures. With the exception of sharks and barracudas, most injuries are a result of either trying to catch game or from contact abrasion with marine life. To treat an injury that results from a dangerous marine animal bite—

- Pack the wound with gauze and then apply a pressure bandage, if available.
- Treat for shock.
- Obtain medical attention as soon as possible.

For a more detailed information on the treatment of casualties and dangerous marine life, see MCRP 3-02F and MCRP 3-02G.

Sharks and Barracudas

Only about 20 percent of all shark species are known to attack people. Sharks have an acute sense of smell, and the smell of blood in the water will draw them to their prey. They are also very sensitive to any abnormal vibrations in the water; therefore, the sound caused by a struggling swimmer or underwater explosions will attract them.

Barracudas are bold and inquisitive fish. They have been known to attack men who are wearing shiny objects. Barracudas may charge at lights or shiny objects at night.

A group of swimmers can maintain a 360 degree watch while in the water. Therefore, to protect yourself from sharks and barracudas, stay with other swimmers. The group members can either frighten away or fight off sharks or barracudas better than one person can. Keep all your clothing on, including footwear. Historically, sharks attack unclothed individuals in groups first, mainly in the feet. Clothing also protects you against abrasions from a shark's tough skin should the shark brush up against you. Avoid urinating heavily, let urine dissipate between discharges. If you must defecate, do so in small amounts and toss it as far away from the group as possible. Do the same if you must vomit.

If attacked, the use of firearms by swim sentries should be used with extreme caution because of the risk of injury to other swimmers. If unarmed or unable to make an improvised weapon, kick and strike the shark. Avoid using your bare hands to strike the shark, injury can result to your hands due to a shark's tough skin. Target areas on a shark are the gills, eyes, and underbelly. Blows to the snout are also not recommended because a shark will tilt its head up and thrust its jaws forward when biting.

Sea Snakes

Sea snakes are venomous and sometimes found in mid-ocean. They are unlikely to bite unless provoked. Your best protective measure is to avoid them.

Poisonous Fish

Many reef fish are toxic, and can kill you if eaten. Generally, poisons are present in all parts of the fish, but especially concentrated in the liver, intestines, and eggs.

Turtles and Moray Eels

Turtles and moray eels normally inflict minor bite wounds. Treat this type of injury by cleaning the wound.

Corals

Coral, dead or alive, can inflict painful cuts. Clean all coral cuts thoroughly. DO NOT use iodine to disinfect any coral cuts. Some coral polyps feed on iodine and may grow inside the flesh.

Jellyfish, Portuguese Man-of-War, Anemones, and Others

This group of marine animals inflicts injury by stinging their victims with their tentacles. Contact with their tentacles produces burning pain, a rash, and small hemorrhages on the skin. Shock, muscular cramping, nausea, vomiting, and respiratory distress may also occur. Gently remove the clinging tentacles with a towel and wash or treat the area. Use diluted ammonia or alcohol and talcum powder to treat the injury if available.

Spiny Fish, Urchins, Stingrays, and Cone Shells

These animals inject their venom by puncturing the skin with their spines. General signs and symptoms include swelling, nausea, vomiting, generalized cramps, diarrhea, muscular paralysis, and shock. Deaths are rare. Treatment consists of soaking wounds in hot water, if available, to deactivate heat-sensitive toxins.

Chapter 4
Negotiating Water Obstacles

Marines face water obstacles in saltwater, freshwater, and brackish water (where freshwater and saltwater meet). These water environments differ considerably and pose distinct problems for Marine tactical units and swimmers. Saltwater obstacles include tides, surfs, and currents. Brackish water obstacles include back bays. Freshwater obstacles include rivers and canals.

Tides

Tides are periodic changes in the surface levels of oceans, bays, gulfs, inlets, and rivers. The Moon's and Sun's gravitational pulls cause tides. Depending on the situation, tides can either help or hinder Marines in their endeavors to conduct amphibious or riverine operations. Tides can provide sufficient depths to allow for the passage of landing craft or boats over reefs, trees, and other underwater obstructions. On the other hand, tides can render a river fording site unusable to tactical vehicles. Direction, level of change, and amount of change determine tidal nomenclature.

Tides that show change in direction are flood tides and ebb tides. Rising tides are known as flood tides. Falling tides are known as ebb tides.

Tides that show extreme levels of change are high tides and low tides. High tide is the period when water is at its greatest depth. Low tide is the period when water is at its most shallow depth.

Tides that show the least and most amount of change are neap tides and spring tides, respectively. Neap tides have the least amount of change in water levels between high and low tide. Neap tides occur at the half moon, when the Sun and the Moon are aligned at a 90-degree angle with the Earth. In this position, the Sun's and Moon's gravitational pulls offset each other. Spring tides have the highest floods and lowest ebbs. Spring tides occur at or shortly after the new moon or full moon when the Sun, Moon, and Earth are approximately in line. In this position, the Sun's and Moon's gravitational pulls are combined.

Surf

Waves break upon entering shallow water and create surf. The offshore area where waves break is the surf zone, which can present many hazards. Breaking waves often trap air bubbles and create a foamy appearance. Bubbles lower the water's density and decrease buoyancy. Move through foamy surf as quickly as possible. The type of wave determines your survival technique.

Wave action moves you toward shore. Lie on your back or side with your head pointing in the direction of the beach and your feet pointing into the waves.

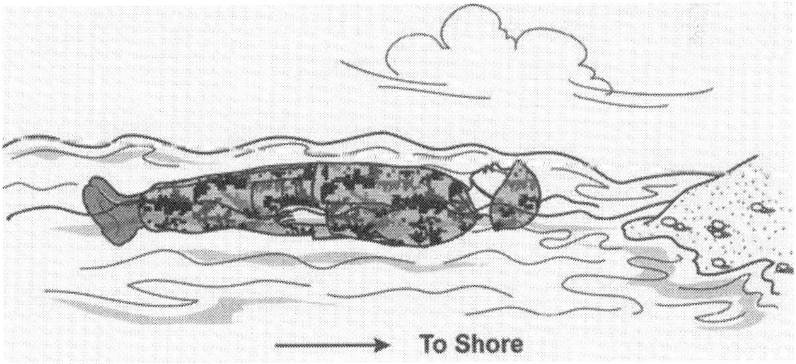

As one wave approaches the beach, another drains away; relax and do not swim against the draining water. When a new wave is within about 3 yards, start swimming toward shore. Continue to swim until the wave lifts you and moves you toward the beach. Once the wave loses forward momentum, relax and repeat the cycle. If nearing rocks, turn your body and approach feet first to reduce the chance of striking your head and arms.

Plunging Waves

A plunging wave is a breaker whose crest curls forward and falls ahead of its base. Because of its power and underwater turbulence, a plunging wave poses the greatest surf threat. If caught in a plunging wave, you can be pulled underwater and pitched about violently. This can cause you to panic, which can increase your chance of drowning. Perform the following steps to escape a plunging wave:

Tuck into a ball with your head against your knees and your forearms locked around your legs, just below the knees.

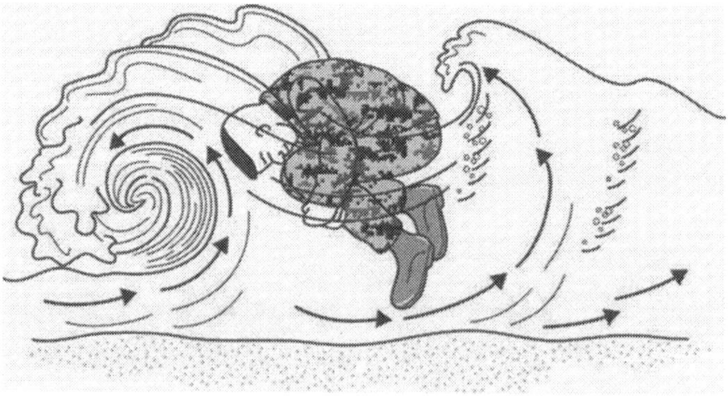

Relax in this position until the turbulence subsides and you float to the surface. This can take 30 seconds or more.

Swim toward shore.

> NOTE: If threatened by another plunging wave, dive underwater into the wave.

Spilling Waves

A spilling wave does not break. Instead, its crest slides forward without curling. A spilling wave creates less turbulence and poses less of a threat than a plunging wave. If caught in a spilling wave, relax and let the wave carry you to shore.

Surging Waves

A surging wave occurs on a beach with a steep underwater gradient. It never really breaks, but the crest rises while the base slides up the beach with great force and speed. Once the wave reaches its highest point on the beach, it rushes back as quickly as it surged forward. If you are standing on the bottom when a surging wave advances or retreats, the wave can knock you off your feet and pull you into the surf zone. Do not try to stand or walk on the bottom. Swim toward the beach as soon as possible.

Currents

Offshore Currents

An offshore current occurs outside the surf zone. Typically, it occurs at bay entrances, in island channels, and between islands and the mainland. An offshore current flows parallel to or away from shore. If the offshore current is created by tides, its current strength and direction vary at different times of the day.

If caught in an offshore current, you may be carried in a direction you do not want to go. DO NOT try to swim directly to safety. If the current is moving directly away from the shore, relax and wait until the current dies out or turns toward land. Once the current subsides, use any survival stroke to swim toward shore. If the current is moving parallel to shore, use any stroke to move at an angle across the current and toward shore.

Rip Currents

A rip current occurs when waves pile water against the shore faster than the water can drain. The water flows rapidly along the beach until it is deflected seaward by a bottom obstruction. Then the rip current flows through the surf zone and into open water at a speed of up to 2 knots. This action can cut deep trenches in the sand. A rip current dies out once it reaches open water (usually within a few hundred yards of the shore).

A rip current can pull you out to the open sea. If caught in a rip current, DO NOT try to swim against the current. A rip current moves faster than most people swim, and it is impossible to swim to shore once caught in it. Relax and stay afloat until the current runs out. Once the current subsides, use a survival stroke to move parallel to the shore until you are out of the current, then begin swimming toward shore.

Littoral Currents

A littoral current occurs when a wave breaks against a beach at an angle. This current flows parallel to the shoreline and does not pose a great threat. If caught in a littoral current, use the combat travel stroke to swim across it at an oblique angle.

Back Bays

Once on the beach, you often face one or more rows of low hills called dunes. Behind the dunes, you may encounter a low-lying stretch of ground thickly covered with scrub trees and bushes. This area gives way to wetlands known as back bays. Back bays consist of muddy islands that are almost submerged during flood tides and separated by channels of brackish water of varying depths. Channel bottoms usually contain soft mud. Back bays pose major obstacles to vehicular traffic.

Infantry can cross back bays, but only with great effort. If crossing back bays by foot, consult detailed navigation charts and use the following guidelines to plan your route:

- Avoid water less than waist deep; walking in shallow water or soft mud is extremely tiring.
- Avoid back bay islands; these low-lying islands are usually too muddy to support foot traffic.
- Seek out deep water; floating with a pack is less tiring than walking through shallow water or soft mud.
- Seek out sand, shell, gravel, or stone bottoms; these firmer bottoms generally ease travel and help conserve energy.

Rivers and Canals

A river is a large, natural stream of water that empties into a larger body of water. The slope of the riverbed and the volume of water in the river determine its current.

Canals resemble small rivers or streams in their width and depth, but usually lack any significant current. Climbing out of these waterways can be difficult if the canal is flanked by steep banks.

Chapter 5
Fording Waterways

WARNING

Fords are dangerous. Cross as quickly as possible.

A ford is any site in a river, stream, or canal where the water is shallow enough for troops or vehicles to cross without using flotation devices. Canal bottoms are usually too soft to support fording vehicles, and wading infantry frequently stumble.

The tactical situation dictates the location of the fording site. Seek fords that are protected from enemy observation and that allow for adequate supporting fires. A night fording takes at least one and a half times as long as a daylight fording.

> *NOTE:* In this chapter, the term "river" refers to rivers, streams, and canals.

Selection of a Ford Site

The following table identifies desirable fording characteristics:

Characteristics	Comments
Concealment	The ford hides personnel and vehicle movement from enemy observation.
Accessibility	The ford should have low banks with gentle gradients. This allows a free flow of traffic at both the entrance and the exit.

Characteristics (Continued)	Comments (Continued)
Slow Current	The ford's current should not exceed 1.5 meters per second if possible.
Firm Footing	The ford's bottom, entry, and exit composition should be firm enough to support traffic. Do not drive a vehicle over any bottom composition that a 2-inch diameter stick can be pressed into more than 1 or 2 inches.
Gently Sloped Channels	The ford's entry and exit points should be gently sloped. If possible, locate a portion of the stream where the channel is not actively shifting.
Depth	The fording depth is less than or equal to the least capable vehicle.

Determine the Slope

Units move into and out of water faster and more quietly if entry and exit points are not steep or muddy. Slope is the amount of change in ground horizontal distance (run) and in vertical elevation (rise) from one point to another. Slope is usually expressed as a percentage. You can use a clinometer, map, or line of sight and pace to measure the percentage of a slope.

Clinometer

A clinometer measures percentage of a slope. It is a component of the M9 armored combat excavator, and organic to many Marine engineer units. However, engineers are also taught (and often use) field-expedient methods to determine slopes and gradients.

Map

A map measures the horizontal distance along a desired path. Determine the difference in elevation between the path's starting and ending points. From a map's scale, you can determine the distance between two points. From a map's elevation lines, you can determine the difference in height between the same two points. Both figures must be in the same unit of measure (e.g., feet, meter). Divide the elevation (rise) by the distance (run) and multiply by 100.

Rise = 165 - 120 = 45 meters

Run = 200 meters

% Slope = $\frac{45}{200}$ x 100 - 22.5%

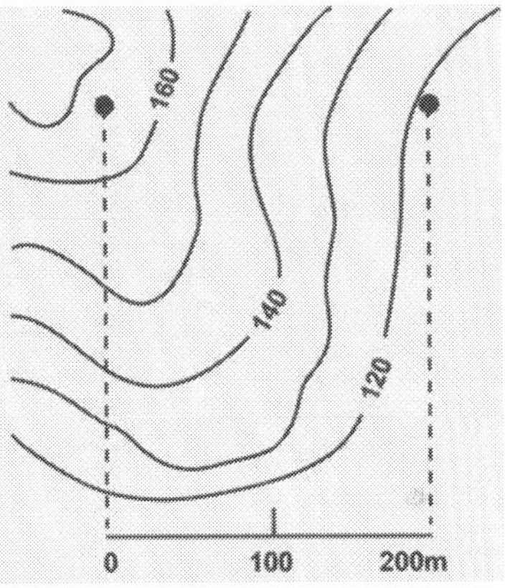

Line of Sight and Pace

To determine line of site and pace, stand at the bottom of the slope, keep your eyes level, pick a spot on the slope, then pace the distance. The number of paces multiplied by a standard measure of 0.75 meter determines the run. The eye-level height (usually 1.5 to 1.7 meters) determines the rise. Repeat this procedure until you have covered the entire distance you want to measure for each spot (vertical and horizontal). Add the vertical distances to provide total rise and the horizontal distances to provide total run.

Determine the Current Speed

Current speed increases as channels narrow. It may be necessary to locate a wider ford location to obtain a slower stream current. The following steps are used to calculate the speed of the current:

Determine points A and B along a channel. Then measure the distance between those two points.

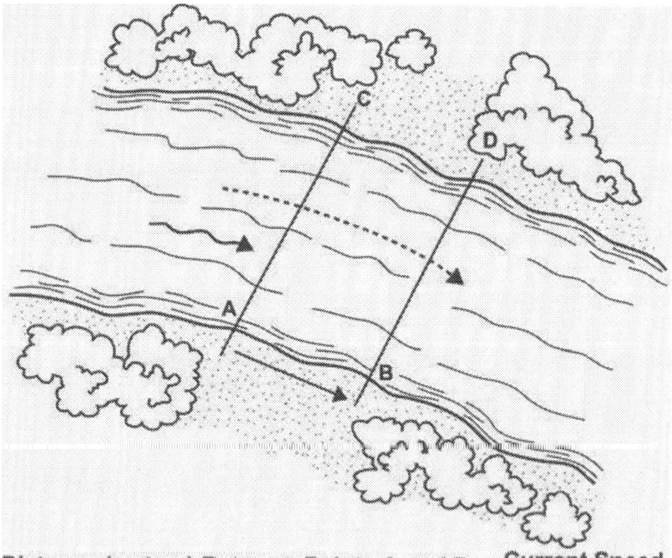

Sight directly across the water from points A and B to locate points C and D.

Throw a floating object (e.g., a stick) upstream from points A and C. Observe the object as it floats toward points B and D.

Subtract to find the time it takes for the object to float from start to finish.

Do not attempt to swim across currents that are moving faster than 1.5 meters per second. Equivalents of this speed include—

- Quick-time march rate of 120 counts per minute with one, 30-inch step at each count.
- 5 feet per second.
- 3.5 miles per hour.
- 5.5 kilometers per hour.

Measure River Width

A river's width can be estimated from the width of its symbol on a scaled topographic map. If this is not possible, use the following compass techniques:

Stand at the water line (A).

Shoot an azimuth to a point on the opposite bank (B).

Move upstream or downstream until you are at a point (C) where you can shoot an azimuth 45 degrees larger or smaller than the original azimuth.

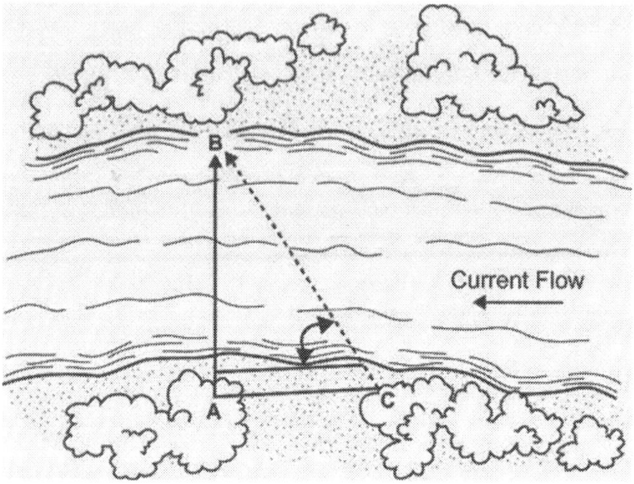

Measure the distance between points A and C. The distance calculated equals the river's width.

Calculate Downstream Drift

A river's current causes personnel and equipment to drift downstream. If personnel and equipment are aimed straight across the river, they will sideslip downstream as they move across the current to the other shore. Therefore, personnel and equipment crossings must compensate for the effects of a river's current and entries are usually made upstream of the desired exit point. Use the formula below to calculate downstream drift.

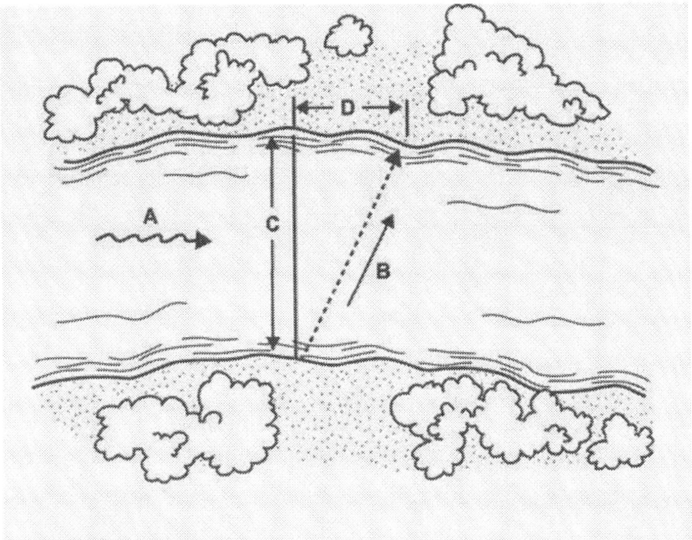

$$\frac{\text{Current Speed (A)}}{\text{Crossing Speed (B)}} \times \text{River Width (C)} = \text{Downstream Drift (D)}$$

Note: The crossing speed for a swimmer across a river may vary but is generally limited to 1 meter per second. All measurements must be in the same unit of measure (e.g., meters, feet).

The Buddy System

Whenever a Marine unit must enter into or operate on the water, a "buddy system" is employed in which every Marine is paired with a swimming partner. The buddy system matches an experienced swimmer with a weak swimmer. The experienced swimmer assists and encourages the weaker swimmer and bolsters confidence during night crossings. If a unit has an odd number of Marines, place the extra person with another pair to form a three-person team.

Water Crossings

Care of Weapons

Marine infantry weapons and munitions are designed to be able to operate after immersion. However, protect your weapons from moisture whenever possible.

A gas-operated weapon can malfunction if water travels down the barrel and enters the gas tube. To protect the gas tube—

Close the weapon's bolt before entering the water.

Seal the muzzle with a condom, balloon, plastic spoon wrapper, or other form of waterproof material.

Tie or melt the protective cover to create a watertight seal.

When the muzzle's protective cover is no longer needed, remove it. Open the bolt and inspect the barrel. If the tactical situation permits, swab excess moisture from the barrel. Test fire automatic weapons, if possible. Field strip and clean weapons as soon as possible. If time does not allow for a complete inspection, rinse inaccessible areas with small amounts of diesel fuel, then dry.

ISOMAT Raft

Construction of an ISOMAT raft is time-consuming. This type of raft should not be employed as part of an attack, but used for logistical purposes (e.g., evacuating stretcher cases, transporting supplies). Use the following steps to build an ISOMAT raft:

Wrap ISOMAT sleeping pads around sturdy sticks.

Use parachute cord and square knots to tie the pads securely in place and to lash stick ends together in a rectangle.

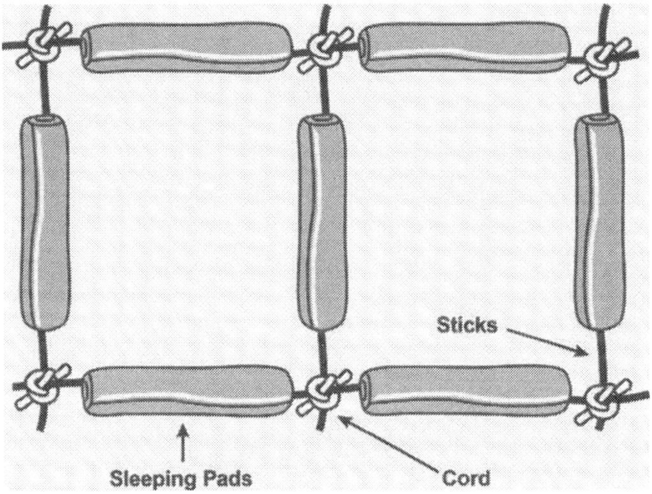

Note: The ISOMAT raft pictured can support several hundred pounds. However, the cargo will get wet if not properly waterproofed.

Poncho Raft

A poncho raft can support two Marines and their equipment and is well suited for long crossings. Use the following steps to build a poncho raft:

Inspect two ponchos and ensure they are serviceable.

Lay one poncho flat on the ground, with the hood-side up.

Cinch the hood tightly to form a gooseneck or tie in a knot.

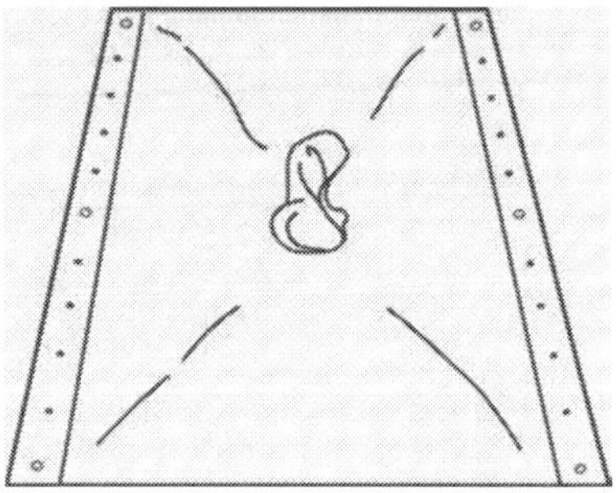

Pad sharp edges of equipment and place the equipment in the center of the poncho.

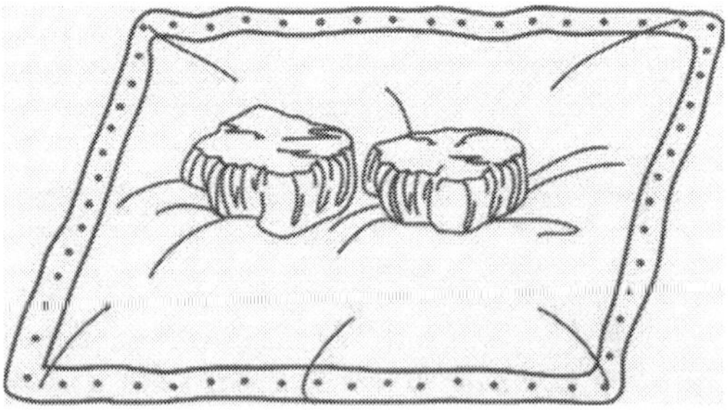

Place the second poncho over the equipment, rubber side up, and hood facing down.

Snap the edges of the two ponchos together.

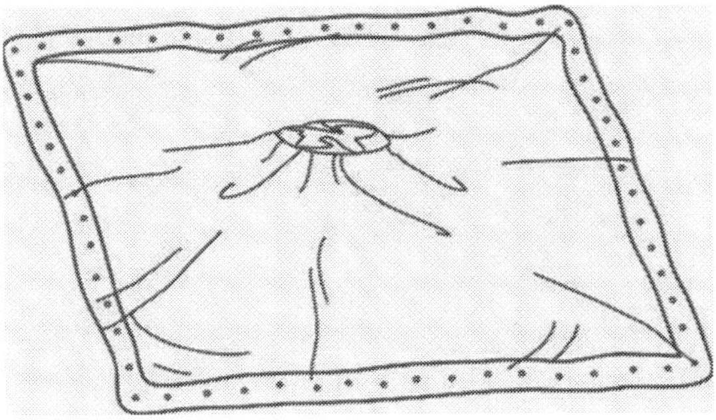

Roll the edges toward the equipment.

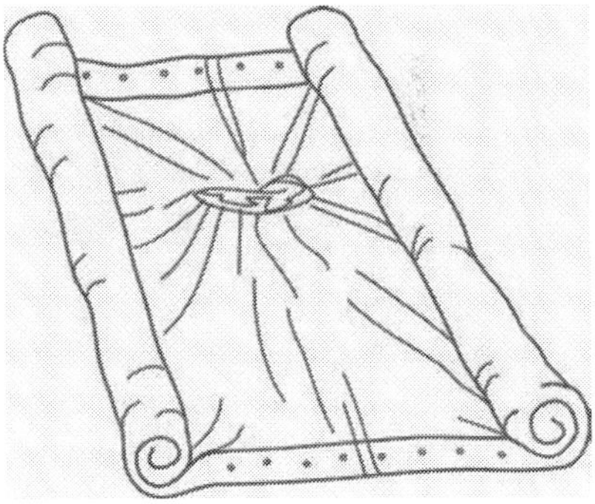

Roll the edges into pigtails and tie them off.

Pull the pigtails together over the top and lash them securely.

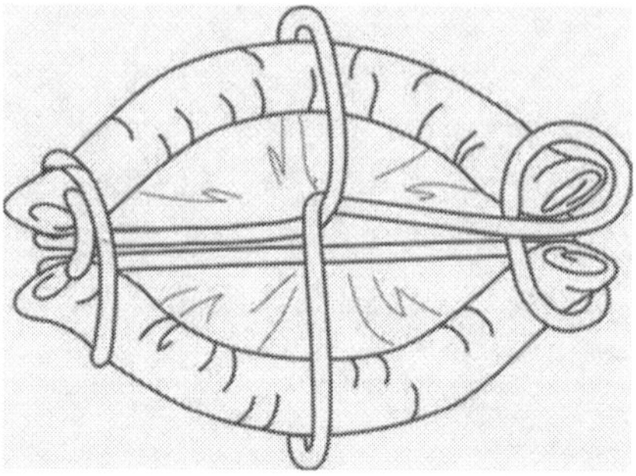

Protect the raft from brush punctures while placing it in the water. Swim across the water obstacle while security elements are covering the far shore.

Construction of a Pack Raft

You will need two waterproofed packs and two M16A2 service rifles. The following steps are required to construct a pack raft:

Place two packs side-by-side with the pack frames on the deck. The tops of the packs are opposite of each other.

Loosen the main compartment straps on both packs.

Insert one rifle on each end between the straps and the packs. The muzzles are opposite of each other. The rifles serve as one means to secure the packs together. Place the front sight post under the top flap.

Tighten the straps so that the rifles and packs are secure.

Take the excess strap on the inner side of each pack and secure it to the opposite pack to better secure the two packs together.

Take the excess straps on the outer sides of the packs and use those straps as safety lashing for the rifles.

Tuck the excess straps and check to make sure the rifles and packs are secure.

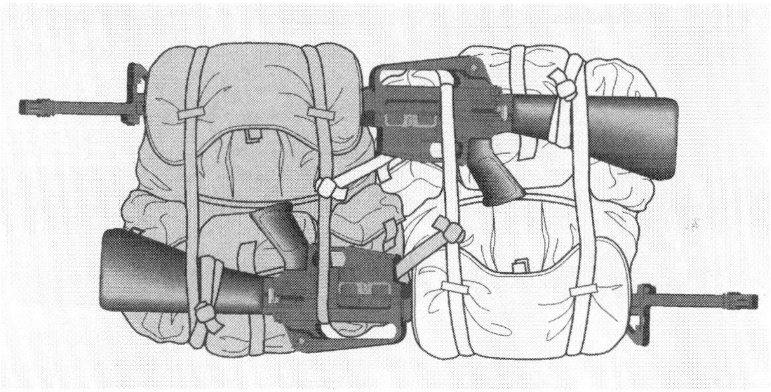

Single-Rope Bridge

A single-rope bridge offers a temporary and quick way to cross small rivers. It also provides extra security while crossing swift waters. At night, it prevents straggling, and guides units precisely from one side of the river to the other side. If crossing a river at night, plan for at least one single-rope bridge.

If your unit is crossing a river with swift currents or water depths above 4 feet, the unit is carrying sufficient rope to span the crossing site, and the tactical situation permits, secure the rope on near and far banks to provide a hand-hold for crossing Marines. This reduces the time required for the entire unit to cross and provides a degree of comfort/confidence for poor swimmers. Use a squad-sized bridge team to construct a single-rope bridge. Station several strong swimmers at the water's edge to help anyone who has trouble crossing.

Nylon rope is normally coiled in 120 foot lengths. It is 0.6 inches in diameter and has a breaking strength of about 3,840 pounds. Over time, a nylon rope can stretch to as much as one-third more than its original length and stretching weakens the rope. If the rope is stretched, discard the rope or use it for light tasks. To prolong the life of a nylon rope, do not step on it or drag it on the ground. Pad the rope in places where it contacts rocks or sharp corners. Do not leave the rope knotted or stretched longer than necessary. Dry rope as soon as possible. Single-rope bridge construction is as follows:

Tie a sling rope around your waist using a square knot and two, separate half hitches. See the appendix for detailed information on knots.

Attach a locking steel carabiner to the sling rope.

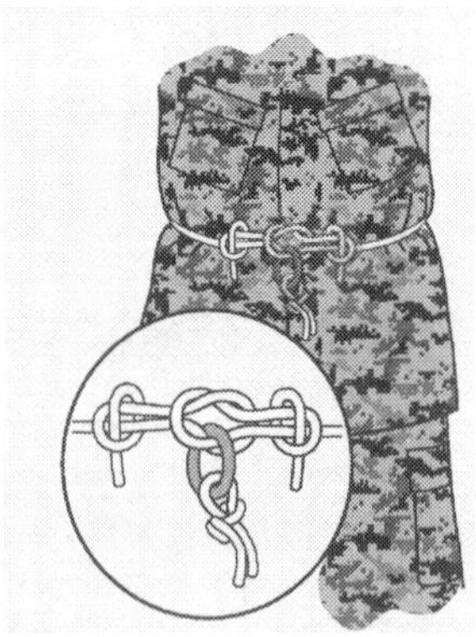

Tie a bowline knot in the running end of the bridge rope and attach it to the carabiner.

Temporarily secure the other end of the rope to a tree on the near shore.

Enter and cross the water.

NOTE: Carry only your weapon and ammunition.

Exit the water on the opposite shore.

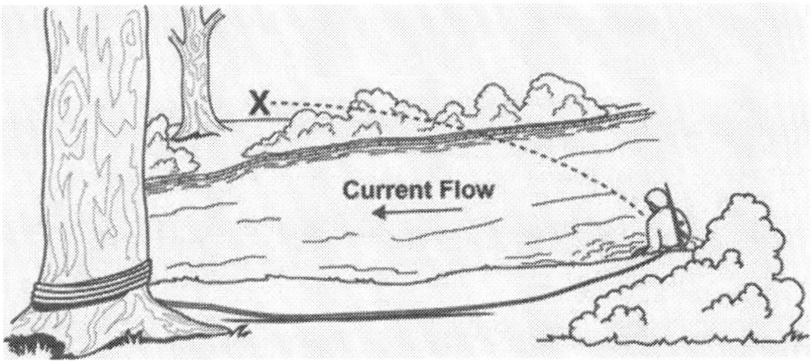

Prepare your weapon for use. Unhook the bridge rope from the carabiner at your waist, and tie the bridge rope to a sturdy tree using a round turn and two half hitches.

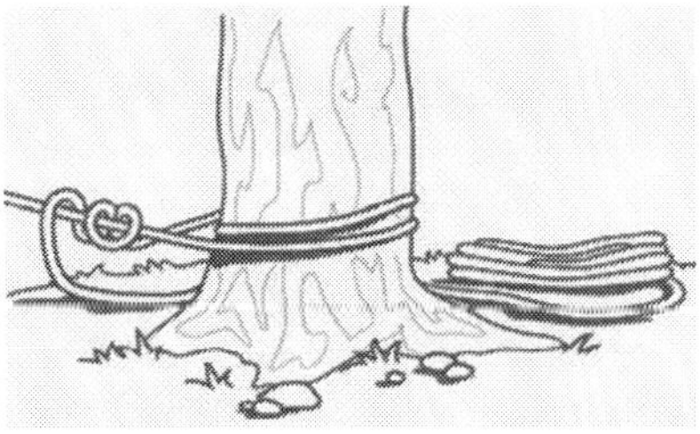

Conduct a box reconnaissance of the opposite shore.

On the near shore, have another Marine prepare to tighten the rope. That Marine should place a transport tightening system in the bridge rope by tying a double butterfly knot and placing two carabiners in the butterfly. See the appendix for detailed information on knots.

The Marine should pass the running end of the bridge rope around the downstream side of the near shore anchor point and through the two carabiners.

Pull the butterfly knot approximately one-third of the distance across the river.

Secure the bridge rope to an anchor point using a round turn and two half-hitches.

On the near shore, the Marine helping you should pull the slack out of the bridge rope until the butterfly knot is back on the near side. The bridge rope is then tied off against itself using two half hitches with a quick release in the last half hitch.

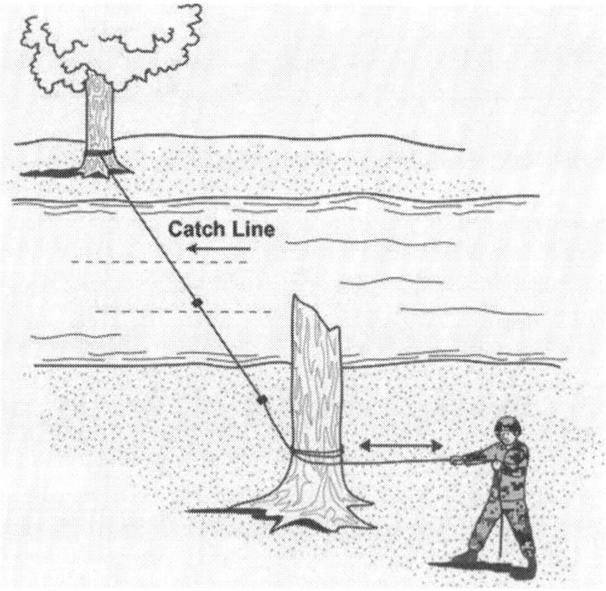

NOTE: The single-rope bridge must be as tight as possible so it will not sag when used.

If you lose your footing and fall into the water, swim with the current to the closest shore. Swimming against the current is dangerous and quickly causes fatigue.

High and Dry Crossings

If the single-rope bridge is high enough, suspend yourself below the single-rope bridge and above the water. Use the following steps to suspend yourself from a single-rope bridge and then pull yourself across the water:

Tie a sling rope around your waist using a bowline. Ensure that the knot is tight.

Attach a carabiner through the bowline's loop. The carabiner's gate faces up.

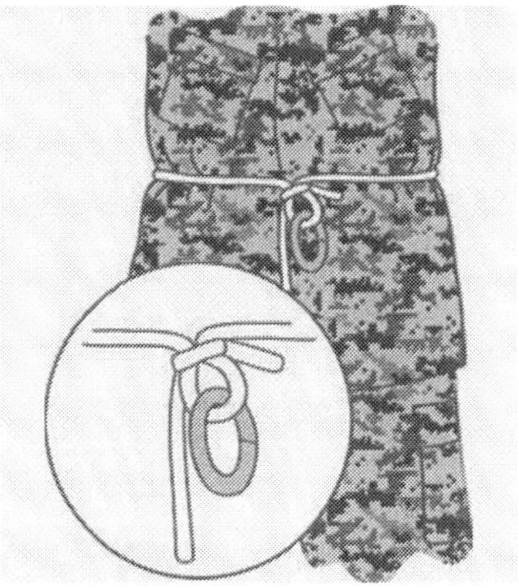

Secure your helmet chin strap.

Face the single-rope bridge with your left shoulder toward the far shore.

Grasp the bridge rope in both hands.

Swing your body beneath the single-rope bridge with your head toward the far shore. Cross your ankles above the bridge rope.

Arch back until the carabiner contacts the bridge rope. Connect the carabiner to the bridge rope. Allow the carabiner to bear your body's weight.

Pull yourself across the single-rope bridge, hand over hand, to the far shore.

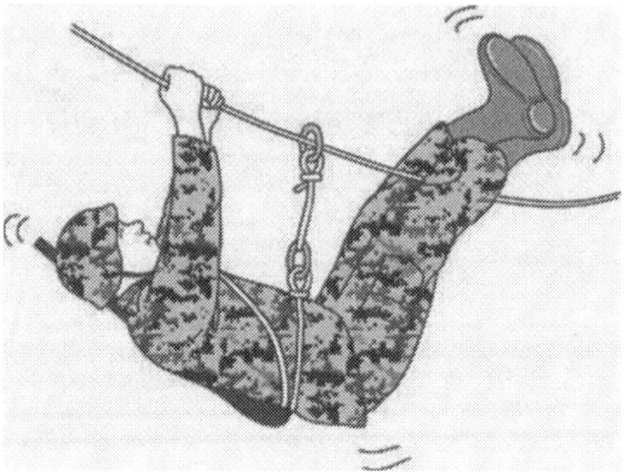

Swift Current Crossings

A single-rope bridge prevents being knocked down and swept away by a swift current. Use the following steps to move through a swift current:

Tie one end of a sling rope around your waist using a bowline.

Tie the running end of the sling rope in another bowline, and attach a carabiner to the bowline's loop.

Step up to the bridge. Face upstream.

Hook the carabiner to the single-rope bridge.

Walk sideways into the river while grasping the bridge rope in both hands.

Use the single-rope bridge for balance and remain standing, if possible.

Continue to move sideways through the river to the far shore.

Slow Current Crossings

If you face little or no current, it is not necessary to hook up to a bridge rope with a carabiner. Lie on your back in the water beneath the single-rope bridge. Support your body weight with your waterproof pack. Use the bridge rope and pull yourself hand over hand across the river.

Removal

If you are the last Marine waiting to cross, pull on the standing end of the rope to release the knot, then tie the rope around your waist using a bowline. The Marines on the far shore will pull you through the water.

Appendix
Knot Tying

Square Knot

The square knot is used to secure two ropes of equal diameter together so they form one continuous rope that will not slip.

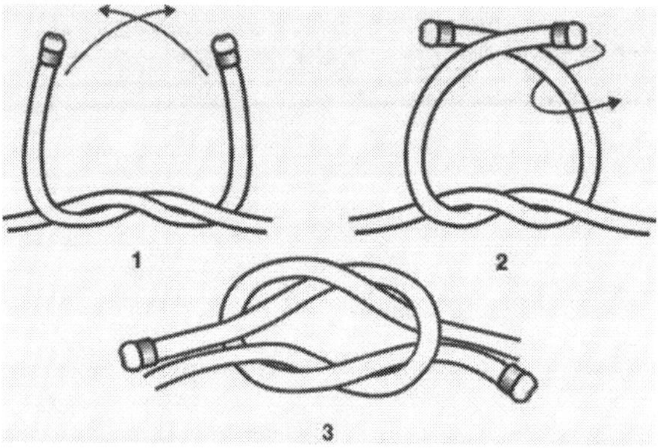

Bowline

The bowline forms a loop that will not tighten or slip under strain. It is easily untied.

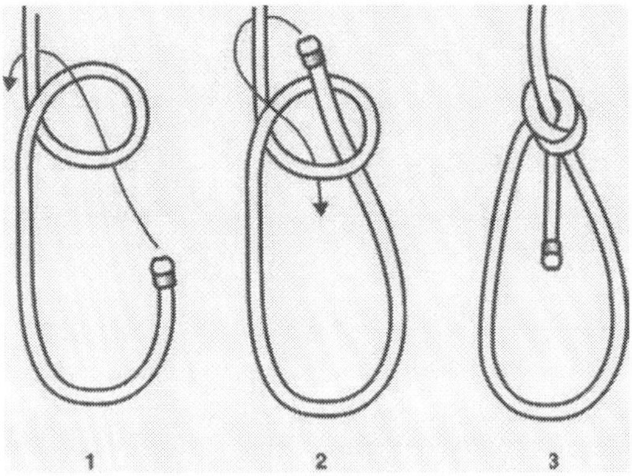

Hitches

Hitches are used to dress (prepare) knots and secure loose ends.

Half Hitch

The half hitch is used to tie a rope to a tree or to a larger rope. It will hold against a steady pull, but is not a secure hitch. It is frequently used to secure the free end of a rope.

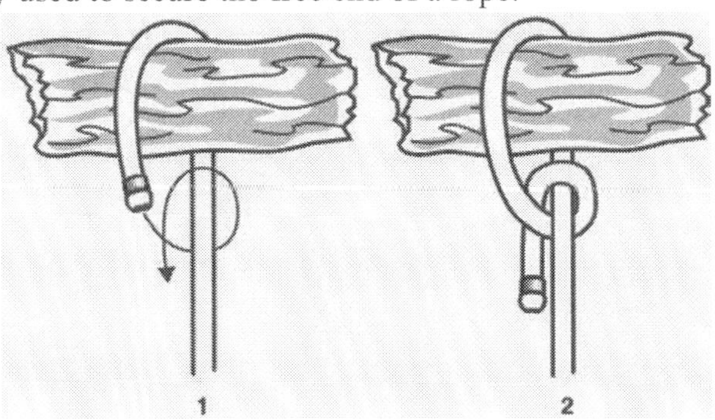

Two Half Hitches

Two half hitches can be used to secure the running end of a rope.

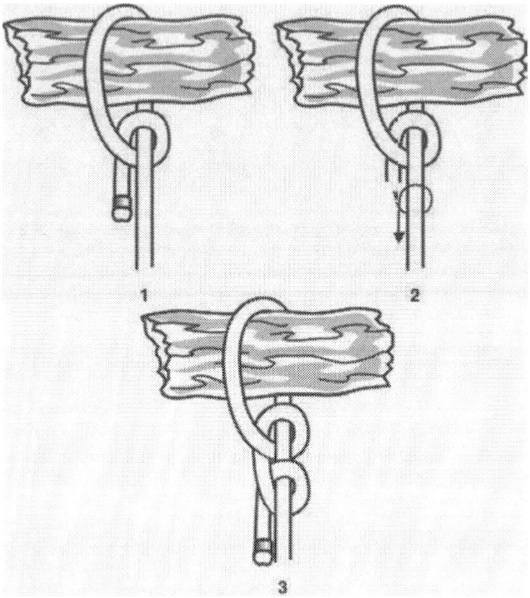

Round Turn and Two Half Hitches

A round turn and two half hitches can be used to fasten a rope to a tree. This hitch does not jam.

Butterfly Knot

The butterfly knot is used for anchor lines. Tied properly, it will not tighten on itself to the point that it cannot be easily untied. The butterfly knot can be used to pull a rope bridge taut. This knot can be used to tighten a fixed rope when mechanical means are not available. It will not jam if a stick is placed between the two upper loops.

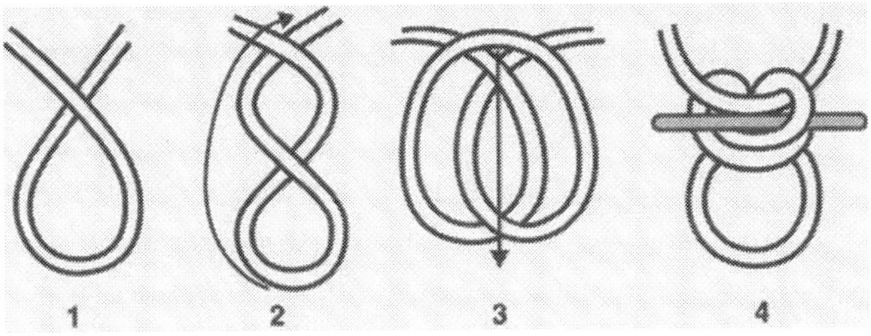

U.S. Marine Corps

Afghanistan An Introduction to the Country and People

Department of Defense

Pentagon
Publishing

Introduction

Purpose

With United States involvement in Afghanistan, knowledge of the country is imperative. This "handbook" explains a number of basic issues that should be in the knowledge "toolbox" of a Marine working with Afghanis or deployed to the country.

Three Forces

Three main forces have shaped present day Afghanistan: Tribalism, Islam, and geography. They have permanent and far-reaching effects on Afghan society and on anyone involved with that society.

Geography

Even though it is a landlocked country, Afghanistan has been on the crossroads of history. The country and people have been at the mercy of numerous empires and conquering states. Afghanistan was a prize worth possessing for its river basins and fertile valley and for its location on the east-west trade routes, especially the Silk Route to China. This geographic factor has affected and changed the peoples of Afghanistan.

Ethnic Interaction

The interaction of the various ethnic groups that invaded the region contributed to the patchwork of people that make up present-day Afghanistan. Each of these external groups that invaded affected the inhabitants of what is now Afghanistan in unique ways.

- For example, Genghis Khan destroyed most of the towns in the thirteenth century, forcing much of population into nomadism.

- A later Turkic dynasty, The Timurids, reconstituted the empire, but was forced out by pressure from newer groups on the west, east, and north of the country.

This constant shift and pressure of various external and internal groups has molded the Afghan attitude toward outsiders and indigenous groups of the country.

Continued on next page

Tribalism

Tribalism is another key factor that has affected the social organization of Afghan society. It affects kinship, loyalties, honor, and even revenge.

[] The family as a unit can reach upwards of 50 people all united by the bond of a common "father."

[] This tribalism, with emphasis on the male or father's family, means that sons often bring their brides to live in their father's home.

Afghanistan remains a patrilineal society, in which membership in both family and lineage, as well as property, is inherited only by kin on the father's side of the family. Kin on the father's side of the family also form close alliances.

International "Tribalism"

This tribalism is not just a local issue. Tribalism is what holds much of the Osama bin Laden's Al Qaida (Al Qaeda) terrorist network together.

[] The bonding and the rules of social relations established in the village and nomadic setting play the same part in the international network.

[] This tribal set of values that applies in the farming and nomadic settlements of Afghanistan and Pakistan has the same validity in Bosnia, Macedonia, and the United States.

Islam

Islam is the last influence that affects Afghanistan in a permanent way. It is a basis for the daily living and for understanding how Afghanis approach interpersonal relationships. It is a guide to the moral and ethical values that Afghanis hold as well as their spiritual mindset.

Continued on next page

Introduction, Continued

Table of Contents

The table below lists the main topics and their page numbers.

Quick Facts on Afghanistan

Introduction	This brief resource guide covers a number of important subjects that give you a clearer understanding of the country. This page of quick facts is like a job aid. It highlights critical issues.
Land of Contrasts	Afghanistan consists of a major mountain range, the Hindu Kush, with its valleys, deserts and river system, http://fermi.jhuapl.edu/maps/afghanistan/ This territory includes many varied geographical and ecological zones. The different peoples of Afghanistan are as distinct and varied as the geography. Bringing them under one flag has made them into a "single" society.
Name	Transitional Islamic State of Afghanistan (TISA) as of December 2002 (formerly called the Islamic Emirate of Afghanistan by the Taliban in 1997) is a land-locked country in southwest Asia.
Size and Population	Area: 652,225 square kilometers or 251, 772 square miles. Estimated population: @ 27,000,000.
Population; Life Expectancy	0-14 years: 42.2%; 15-64 years: 55.01%; 65 years and over: 2.79%. Life expectancy about 46 years.
Languages and Ethnicity	Pushtu 35%, Afghan Persian (Dari) 50%, Turkic languages (primarily Uzbek and Turkmen) 11 %, 30 minor languages 4%.
Religion	Sunni Muslim 84%, Shi'ite (Shi'a) Muslim 15%, other 1%:.
Administrative Sectors	Country is divided into 30 provinces or velayets. The largest in area are Herat, Helmand, Qandahar (Kandahar), and Farah. The largest in terms of population are Kabul, Herat, Balkh, and Ghazni.
Population Centers	Principal towns include Kabul, Kandahar (Qandahar), Herat, Mazar-i-Sharif, Jalalabad, and Ghazni. http://www.afghana.com/GetLocal/Afghanistan/Provinces.htm

Continued on next page

Quick Facts on Afghanistan, Continued

Government/ Legal System	A transitional government, consisting of a grand assembly, president, and judiciary; no longer based on the Qur'an (Koran), the holy book of Islam.
Politics	Taliban (religious students) headed by Mullah Mohammad Omar was deposed in late 2001; a transitional government based on United National Islamic Front for the Salvation of Afghanistan or UNIFSA, often referred to as the Northern Alliance, and exiled leaders, forming a legislative assembly, a presidency, and a judiciary.
Economic Sectors	Agriculture: 53% of gross national product (GNP); industry: 28.5% of GNP; services: 18.5% of GNP (1990).
Additional Sites	http://www.odci.gov/cia/publications/factbook/gcos/af.html — CIA site that has handbooks of countries of the world. http://ebooks.whsmithonline.co.uk/htmldata/encv.asp?mainpage=HTTP://EBOOKS.WHSMITHONLINE.CO.UK/ENCY CLOPEDIA/01/C0000001. HTM — short summary about Afghanistan from *The Hutchinson Family Encyclopedia.* http://www.nutshellnotes.com/afghanistan text.htm - short dictionary-like definition information about the country.

Geography and Climate

Land-locked

Afghanistan is a landlocked country in southwest Asia. To the north, its neighbors are Turkmenistan, Uzbekistan, and Tajikistan. Iran is to the west; the People's Republic of China to the northeast, and Pakistan to the east and south, http://www.lib.utexas.edu/maps/cia01/afghanistan smOl .ipg

Chief Topographic Feature

The Hindu Kush mountains are the main topographic feature that marks Afghanistan. These mountains are a barrier between the northern provinces and the rest of Afghanistan. They divide the country into three distinct geographic areas: the central highlands, the northern plains, and the southwestern plateau. You can find good maps of Afghanistan at http://www.lib.utexas.edu/maps/middle east and asia/afghanistan topo86.jpg

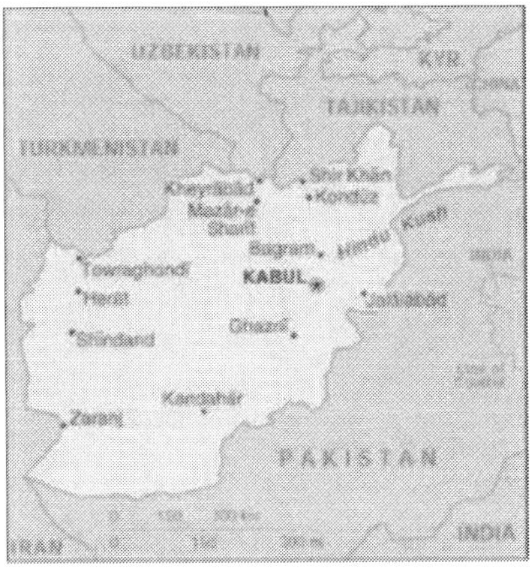

Source: http://www.globalspecops.com/oef.html

Continued on next page

Geography and Climate, Continued

Central Highlands

The central highlands are an area of about 160,000 square miles. This part of Afghanistan has deep, narrow valleys, as well as high mountains, which have proven to be historically important to the defense of the country. The Khyber Pass, one of the most famous routes to the Indian subcontinent, is located in the mountain ranges of the central highlands, http://www.maps.com/cgi-bin/search/hyperseek.cgi?search=CAT&Calegory=Asia%3AAfghanislanP&0ualifier=

The climate of the central highland is usually dry, with temperatures in the summer averaging around 80 degrees Fahrenheit; the winters are very cold.

Southern Plateau

This region of Afghanistan consists of high plateaus and sandy deserts. This desolate region covers about 50,000 square miles and is crossed by several large rivers. The average altitude of the southern plateau is about 3,000 feet. Kandahar (Qandahar), at an elevation of about 3,500 feet, enjoys a dry, yet mild climate.

Northern Plains

The northern plains are about 40,000 square miles of extremely fertile foothills and plains.

- The Amu River (formerly called the Oxus River) runs thought the foothills of this region.

- The average elevation is about 2,000 feet.

Agriculture is the main occupation in this region. There are large amounts of mineral deposits and natural gas in the northern plains.

Climate

The climate of Afghanistan varies dramatically between the mountain country and the lowlands. In the southwest during summer the temperature can reach 120 degrees Fahrenheit. In the Hindu Kush mountains during winter it can fall as low as -15 degrees Fahrenheit

Map Site

http://www.lib.utexas.edu/maps/afghanistan.html — This site has many different kinds of maps of Afghanistan.

Economy and Livelihood

Economic Areas

[] Most people in Afghanistan are farmers or peasants who cultivate crops for personal use or for marketing and selling as produce. Herding is the next favored occupation by population size, followed by forestry and fishing.

[] Manufacturing is the second largest area, followed by wholesale and retail commerce, then by transportation and mining. (With the current chaotic situation in the country, much of the economy has been disrupted.)

Agricultural Economy

Agriculture requires five major means of production: land, water, seed, labor, and animals (or tractors). Water is the most important resource in this respect. Each village has an official chosen by the elders or the village council. He regulates the division of water among landowners.

Family Farming

The family farm is the most common form of ownership and land cultivation. The family usually provides the labor and animals to cultivate its own land. Other agricultural labor arrangements include:

[] Sharecropping, where the tenant receives a part of the produce.
[] Commercial fanning, where the owner pays cash to the farmers.

Because of the current chaotic conditions in Afghanistan, these agricultural systems are in decline.

Herding

Pastoral nomadism has been the traditional way that Afghanis have raised and herded their cattle, sheep, and goats. These nomads usually maintain their herds in groups, traditionally as clans or tribes. They might rely on the peasants or farmers for staples such as grain and implements they need for running a household.

Continued on next page

Mobility: An Advantage

Because of their mobility nomads were:

[] The bearers of information and news.

[] A traditional source of credit because their herds were a "countable" tangible source of cash. In the towns and villages where they took their stock to sell, nomads could buy goods on credit to sell to the isolated farmers or peasants living near their summer pastures.

[] Often smugglers, both inside the country and across the borders.

Again, based on the chaotic war conditions of the country the usual nomadic herding patterns may have diminished or disappeared.

Manufacturing and Commerce

Besides food staples, such as margarine and wheat flour, woven fabrics, clothes and shoes are produced.

[] Fertilizer, cement, and electric energy are the three main industrial products.

[] Mining includes hard coal, gypsum, and natural gas.

The war in Afghanistan has caused industrial and commercial output to decline.

Languages and Ethnicity: Diversity

Introduction

As a landlocked nation, Afghanistan is at the crossroads for much migration and many settlements. The result of this exposure is a country with diverse ethnic and linguistics (language) groups.

Two Perspectives

[] The next two sections give you an idea of the kinds of people living in Afghanistan, including the ethnic groups of the country. An ethnic group consists of people who have many cultural traits in common. These may include language, religion, morals, customs and ceremonies, economics and livelihood.

[] Another way to look at diversity in a country is according to the languages the peoples of the country speak. Languages may cross ethnic boundaries, for example in Afghanistan, Dari is a common language to many of the different ethnic groups in the country. It crosses the ethnic/cultural boundaries of these different groups.

Languages of Afghanistan

Pushtu (Pashtu) Pushtu is the official language of Afghanistan as, for example, Turkish is the official language of Turkey. Pushtu is also spoken in northern Pakistan. The language is written in a modified Arabic script.

Dari Dari or Afghan Persian or Eastern Farsi is spoken by about 43% of the population. It is often referred to as the language of business and commerce.

Turkic Languages About 11% of the population speaks a Turkic language, mainly Uzbek or Turkmen. Afghanis living close to the border of Turkmenistan and Uzbekistan probably speak Turkic languages.

Ethnic Groups: Pashtun

Introduction

Like the languages of the country, there are various ethnic groups in Afghanistan. The main groups are the Pashtuns (pronounced PAH-shtoons) (Pathans), Uzbeks, Hazaras (Hezoras), Tajiks (Tadjiks), and other minor groups.
http://www.lib.utexas.edu maps/middle east and asia/afghanistan ethnoling 97.jpg

Definition

The Pashtuns live in southern and eastern Afghanistan in a population belt stretching from Kabul in the east to Kandahar (Qandahar) to Zaranj, which is on the border with Iran. They also live in western Pakistan.

[] Pashtuns speak Pushtu (Pashtu) and are Sunni Muslims (see the section on **Islam** for an explanation of Sunnis). Many Pashtuns also speak Dari, the second language of Afghanistan.

[] Pashtuns make up about 38% of the population of Afghanistan. Until the recent Taliban take-over, the Pashtuns controlled the ruling dynasty of the country for about 200 years.

Settlements

Pashtuns generally live in tribal villages. A number of villages may be near a town. A village is built primarily with regard to the location of water and to self-defense. The dominant lineages or extended families usually have the best and most strategic lands. Familial ties determine a group's location relative to the dominant tribe.

Nomads

Some Pashtuns are herders. They move from site to site following the pastures as the seasons change. Like the permanent villages mentioned above, the campsites of these nomadic Pashtuns are based on tribal lineages. Their tents are made from goat hair, supported by posts or arched poles.

Economy and Livelihood

Pashtuns work at grain farming and animal husbandry (see **Nomads** above). Wheat is the most important crop. Other important products are fruits, nuts, vegetables, and opium.

Industry usually includes trades such as shoemaking and carpentry. Some Pashtuns are small businessmen and traders.

Continued on next page

Ethnic Groups: Pashtun, Continued

Kinship

The tribe is the unit of kinship. Descent is claimed from the father's side of the family.

[] The clan is the principal division of a larger tribe. A clan is usually the offspring of one man, often including four or five generations, what is sometimes referred to as an extended family in Western terms. In this clan, a person marries and forms bonds.

[] The smallest kinship unit is a household. The household or *kor* is the main economic and social unit; its members may form a village, single housing compound, or a nomadic group.

Local Authority

Pashtuns have a concept and practice of local authority based on tribalism.

[] Honor *(nang)* is a tradition that pervades Pashtun society. It means there is little social class and no central authority.

[] Taxes or rent *(qualang)* gives status and social class to the local people, many of whom are landowners. Qualang also means that there may be a "centralized" authority at the local level.

National Government

Generally, villages consider that the national government has responsibility for conscripting young men into the military and for taxation, both of which take resources away from the clan/tribe. Thus, villagers may have an attitude of neutrality toward, non-cooperation with, or defensiveness against the central government.

Important Customs

Pashtuns generally believe in and follow these customs:

[] Purdah: Strict separation of the sexes in public
[] Chadori: Women wearing the veil.
[] Pakhtunwal: A mixture of blood-line revenge, hospitality to guests, defense of those who are in one's care, chastity of married women and caring for the weak

Ethnic Groups: Tajik (Tadjik)

Definition

Tajiks (pronounced TAH-jecks) make up about 25% of the population of Afghanistan. They live

[] In a large area north and west of Kabul (this area borders on the Republic of Tajikistan)
[] Clustered around the city of Herat in western Afghanistan

Tajiks are also the principal inhabitants of the Republic of Tajikistan, a former region of the Soviet Union.
hltp://news.bbc.co.uk hi english static in depth world 2001 war on terror key maps ethnic groups.stm

Language and Culture

Tajiks are primarily Sunni Muslims. They speak the Tajik and Dan languages. Tajiks are believed to be descendants of the ancient Persians.

Economy and Livelihood

Since Tajiks in urban areas are well educated and wealthy, they make up the elite of Afghanistan. Their influence is in commerce.

[] The plains-dwelling Tajiks live mainly in Herat, Parwan, and also around Kabul. They are town dwelling and are traders, skilled artisans, and farmers, all middle-class. Because they have settled in towns, they have replaced their tribalism with an urban orientation and a strong sense of community loyalty.

[] Tajiks in rural areas make their living by fanning and herding. They are a sedentary people. Landowners are usually village leaders.

Politics: Importance in Northern Alliance

Tajiks are the second largest group after the Pashtuns. They are also the Pashtuns' closest rivals for power and prestige, but with two exceptions, one in the 14th century and one for 9 months in 1929; the Tajiks have never ruled their region.

Tajiks fonned an important part of the Northern Alliance, which opposed the Taliban. From a group of Tajik farmers, the chief leader of the Northern Alliance, Ahmad Shah Masoud, emerged. First he fought a guerilla war against the Soviet Communists and later against the Taliban. Taliban supporters assassinated him in late summer 2001.

Ethnic Groups: Hazara (Hezora)

Definition

Hazaras (pronounced HAH-zahrahs) make up about 19% of the Afghani population. They are Shi'ite Muslims, not Sunnis (for more details on Shi'ites, see the section entitled **Islam).**
http:;/news.bbc.co.uk hi english static in depth world 2001/war on terror'key maps /ethnic groups.stm

Groups/Settlements

The Hazaras fall into three groups: Settlements are located in an areas bounded

[] On the north by the town of Bamian (Bamyan)
[] On the west by Chaghcharan
[] In the east by Kabul
[] On the southeast by the town of Ghazni

The Hazaras themselves call this area Hazaristan, country of the Hazaras. Many Hazaras were forced off their land by the expansion of the majority Pashtuns and now live in Kabul and other towns.

Language

The Hazaras speak a language called Hazaragi. It is a mixture of Farsi (the language of Iran), Mongol, and Turkish.

Religion

[] Because they are Shi'ites, the Hazaras have been subjected to oppression over the years. Most recently, the Taliban massacred about 300 Hazara men in January 2001. The reason was the Hazaras were suspected of collaborating with the United Front (Northern Alliance).

[] Since they are Shi'ites, the Hazaras get support from Iran, which is predominantly a Shi'ite Muslim country.

http://news.bbc.co.uk/'hi/english uorld south asia newsid 1500000 1500682.stm — discusses Taliban persecution of Hazaras.

Continued on next page

Ethnic Groups: Hazara (Hezora), Continued

Origins

The origin of the Hazaras is interesting. Some researchers believe that Haz.aras are the descendents of the 12th century Mongol warrior, Genghis Khan and his followers; others trace the Hazaras to the Buddhist civilization of 2000 years ago, the Bamiyan, whose center was in Afghanistan.

http: // www .hazara. net/hazara/geography/B uddha/buddha.html — an article about relationship of Hazaras to the Buddhist statues.

Economy and Livelihood

The primary occupations of the Hazaras are fanning and trade. The main crops are wheat, barley, and tobacco. In some of the warmer areas, rice is cultivated.

Hazaras raise sheep, horses, and camels. Their horses are especially bred for hard work in the mountains. The wool from Hazara sheep is considered the best in Afghanistan. Camels are raised for their hides.

Social Organization

Like the Pashtuns, the Hazaras are organized by "tribes" or large clans. There are about 15 of these super-clans. The members belonging to a specific clan usually live in a certain geographic region. With so many clans, historically, there has been much inter-clan rivalry

Tribe/Clan: Examples

- **Daizinjat** clan lives in the area of Badsken, which borders on Herat and Koshak in the east. In Khila there are about 1,000 families. The Daizinjat tribe has about 30 clans. The most powerful is the Kaka clan.
- **Daikundi** consists of five tribes: Dolat Beg, Roshan Beg, Heider Beg, Gawash, and Barat. The greenest land of the Hazaras is in Daikundi, which is also called Hazarastan Kashmir. Before the Hazaras, the Tajiks lived in Daikundi.

Local and Central Governmental Authority

Because of their position as a religious and ethnic minority in Afghanistan, there has been constant tension between the local Hazara tribes/clans and the central authority of the country. The tension has boiled over a number of times into open revolution against the central authority, most notably in the early 1890s.

Ethnic Groups: Uzbek

Definition

The Uzbeks (pronounced OOZ-becks) are a Turkic-speaking ethnic group, which makes up about 6% to 8% of the population of Afghanistan. Most of them are Sunni Muslims. They are also the dominant population in the neighboring country of Uzbekistan.

http: news.hbc.co.uk hi english sialic in depth world 2001 war on terror key maps /ethnic groups.stm

Location

Uzbeks live in a population belt that stretches along the northern border of Afghanistan from Maymaneh through Mazar-e-Sharif and Konduz. This Uzbek population belt parallels Afghanistan's northern border with Turkmenistan and Uzbekistan.

Social Structure

There are two types of Uzbeks: farmers and nomads. The farmers and city dwellers are mainly businessmen and tradesmen. The nomads raise sheep, goats, and cattle.

Settlements

The basic living unit is the village. Again, there is a nomadic village and a more permanent farming/commercial village. Both villages are based on kinship.

[] The nomadic village is mobile, moving to pasturage as the seasons change.
[] The farming village is a concentrated, permanent settlement: Houses are built close together. Most houses are walled and form a compound with a central courtyard.

Importance of Uzbeks in Afghanistan

Although a smaller minority in Afghanistan, the Uzbeks have played an important role in the country. One of the most important and most controversial groups who opposed the Taliban was the forces under the control of the Uzbek leader Abdur Rashid Dostum.

[] Dostum had allied himself with the government in Kabul but later turned against it and joined with the *mujahidin* in 1992.

[] He is currently "at odds" with the present-day government of Afghanistan. http://www.afghan-info.com/Politics/Afghan Mujaheedin/Warlords.htm

Recent Afghan History

Introduction

This section emphasizes recent history of Afghanistan to show the events that led up to the Taliban taking power in the country.

http://nths.newtner.kl2.il.us/academics/social/conflict/4s%20Web%20Pages/Period9/Afghanistan/afghanhome.htm — a more detailed and lengthier history of the country.

Communist Take-over

[] In 1978, during the Saur Revolution, the Afghan communist party under Muhammad Taraki overthrew the republican government. This communist party, called the People's Democratic Party of Afghanistan, started land redistribution, secularized the country and began to re-educate the people.

[] Factionalism caused the party to break down. This allowed the Soviets to invade in 1979. Eighty thousand Soviet troops invaded the country. The Soviets replaced Taraki with their puppet Babrak Kamal. http://www.afghan-info.com/Historv/Polcvent. htm#saour

At this point groups of Islamic tribes/clansmen formed a loose alliance that collectively became known as the *mujahidin.* They were anti-Soviet and anti-Communist.

Soviet Domination/ Rise of *Mujahidin*

During the 10-year occupation of Afghanistan, about five million Afghanis fled the country; close to 1.5 million were killed.

[] During this period, the *mujahidin* gained prominence as the group that fiercely opposed the Soviets. They were supported by the West and by Muslim countries in their struggle against the Soviets and the puppet government that the Soviets had imposed on the country.

[] The Soviets tried a policy of reconciliation with the opposition *mujahidin* in 1987, but the seven-party *mujahidin* alliance refused to observe a cease-fire and to participate in negotiations with the Soviet puppet regime.

Continued on next page

Recent Afghan History, Continued

Soviet Domination/ New Constitution

A new constitution was adopted in July of 1987, allowing the formation of other political parties beside the communists.

[] The communist party of Afghanistan said it would be willing to share power with opposition groups in a coalition government.

[] The name of the country was changed back to the Republic of Afghanistan from the Peoples Democratic Republic of Afghanistan.

Soviet Withdrawal

The Soviets withdrew their troops in 1989. The last president, Najibullah, remained in power. Civil war continued as the *mujahidin* were strongly against the government of President Najibullah.

***Mujahidin* Rule: New Government**

A number of attempts were made to integrate the *mujahidin* in the central government through advisory councils of tribal leaders.

[] Eventually Rabbani was elected president of the country.

[] In 1992, the Peshawar Agreement was ratified: A purely Islamic interim government was installed in place of Najibullah's.

New Rules for Afghan Society

[] Leadership in this new government rotated among the seven Islamic *mujahidin* groups.

[] This coalition government banned alcohol and gambling, required women to wear the veil, and separated education into male and female institutions.

Factionalism

Infighting among the various factions of the *mujahidin* continued, even though each faction had two ministerial posts in the government.

[] Eventually the alliance broke down into civil war. The most serious conflict was between Premier Hekmatyr, a Pashtun, and President Rabbanij a Tajik, together with his defense minister, Ahmad Shah Massoud.

[] The result of this armed conflict was the almost total destruction of Kabul and more important, Pakistan's support for a group of Pashtun religious students, the Taliban.

Taliban's Rise to Power

Early Advances

In September of 1994, the Taliban (meaning students in Persian), an "army" of former Muslim seminarians and religious students, entered the civil war. By 1995, the Taliban had claimed victory in the town of Herat and had advanced on Kabul, the capital.

During 1995, the Taliban made extensive gains in the country, particularly in the south. In many instances, they were welcomed by the local Afghanis who were tired with and disgusted by the chaos, corruption, and instability that mujahidin factionalism had caused.

Occupation of the Capital

[] By autumn of 1996, the Taliban had driven President Rabbani out of Kabul, captured the capital, and executed Najibullah. In May 1997, after a brief alliance between opposition forces and the Taliban, violent civil war again broke out.

[] Fighting then continued between the United National Islamic Front (Northern Alliance), consisting of the *mujahidin* factions, and the Taliban.

Afghanistan divided: Taliban

The Taliban held about 85-90% of the area of Afghanistan. http://europe.cnn.com/SPECIALS/2001/trade.center/afghan.civil.html , http:.7w\\\vl.cnn.com/WORLD/'J61()/(i4alahan/iiidc\.lHiv - good discussion of the problems the Taliban have in uniting the country.

United Front/Northern Alliance

[] The Northern Alliance held the remaining 10% to 15% of the country. As of late November 2001, the alliance had extended its control over the country, capturing Kabul.

[] The alliance was strongly against the Taliban's extreme fundamentalist methods and religious philosophy and was very much opposed to Pashtun ethnic dominance, http://www.washingtonpost.com/wp-srv/nation/graphics/attack/zone 3.html

http://www.afghan-web.com/politics/parties.html - lists anti-Taliban parties/groups and their leaders.

http://www.afghan-web.com/ciriello/alliance/ - pictures of alliance forces.

Taliban: Who Are They?

Education and Training

The Taliban were originally young Afghanis many of whom went to Pakistan to study in madrassas (religious schools). These schools:

[] Focus on instruction in the Qur'an; however, a number of them teach military training and militant Islam to their students.
[] Were the source of the Taliban's interpretation of Islam and their approach to restructuring Afghanistan socially and politically. (Saudi and Pakistani intelligence services also trained these Afghan religious students.)

With their "purist" Islamic view, the Taliban were considered a group that promised a clean, uncorrupt society for Afghanis and were opposed to the factionalism and corruption of the *mujahidin,* who had plunged Afghanistan into civil war after the withdrawal of the Soviets.

Outlook

The Taliban are very provincial and backward in their outlook: The external world, much of which they considered unclean and corrupt, begins at the border of Afghanistan.

The Taliban have a close relationship to Osama bin Laden, whom they feel follows the same fundamentalist precepts they believe in. Their relationship to bin Laden was further cemented by the marriage of their leader, Mullah Omar, to one of bin Laden's daughters.

Taliban Oppression

Besides the requirement for women to be fully veiled and accompanied by a **man** in public, the Taliban had

[] Stopped all education for girls and women
[] Forbidden women to work
[] Forbidden women to wear white socks
[] Required men to grow beards
[] Stopped the Afghan national sport of Buzkash
[] Changed the name of the country to the Emirate of Afghanistan
[] Forced conscription upon the male population of Afghanistan to increase their fundamentalist army

The law of the country was the Taliban interpretation of the Sharia, Islamic religious law, with strict penalties, such as death and stoning, for infringements of the various oppressive laws imposed on the Afghan population.

Taliban Relations with Neighboring Countries

Iran

Tensions between Taliban-controlled Afghanistan and Iran were always high. The Iranians were angered by the murder of their diplomats and a journalist by the Taliban near Mazar-i-Sharif. In response to these killings, in September 1998, Iran deployed 70,000 troops to conduct military exercises near the Afghan border, but no fighting occurred between the two nations.

[] The Iranians also feel a "kinship" to the Shiite Hazaras of Afghanistan, whom the Taliban have persecuted and massacred.

[] Finally, the extreme primiliveness and barbarity of Taliban fundamentalism alarmed Iran. The Iranians were probably glad to see the Taliban removed from Afghanistan as the controlling government authority.

Pakistan

The Taliban were supported among some groups of the Pakistani population and by some Afghan refugees living in Pakistan, particularly those close to the Afghan-Pakislani border and those trained in the militant madrassas. The kind of fundamentalism preached by the Taliban is also prevalent in some sectors of Pakistani society.

Although the government of Pakistan supported the U.S. effort to root out bin Laden and his network of terrorists based in Afghanistan, Pakistan still recognized the Taliban as the legitimate government of Afghanistan long after the rest of the world denied their legitimacy.

Other Countries

Tajikistan, Uzbekistan, and Turkmenistan were all apprehensive about the Taliban, particularly their type of Islamic fundamentalism. Many of these former regions of the Soviet Union, now independent countries, are struggling to establish democratic processes for their populations. They viewed the spread of Taliban Islamic fundamentalism as a challenge to their own move toward popular sovereignty.

On the other hand, since there are minorities from these countries, particularly Uzbeks and Tajiks, living in Afghanistan, the governments of these neighboring countries had to be careful about the amount and intensity of their anti-Taliban support.

Post-9/11 Afghanistan

9/11: Activating the US War on Terrorism

September 11, 2001 is a date that saw the United States and eventually other coalition countries begin a concerted effort to actively combat international terrorism, primarily represented by Osama (Usama) bin Laden. He was eventually placed on the FBI's "Ten Most Wanted List" with a $5 million reward for information leading to his arrest.

After 9/11, the U.S. government requested that the Taliban government of Afghanistan turn over bin Laden, the international terrorist who had masterminded the terrorist attacks on the World Trade Center (WTC) and the Pentagon.

[] When the Taliban refused to extradite bin Laden, a US-led coalition was formed and after six weeks of aerial bombardments and military action on the ground the Taliban was ousted from power on November 17, 2001.

[] Many Taliban dispersed to mountain retreats in the country or fled across the border into neighboring Pakistan. Osama bin Laden is still at large and his whereabouts are not known. The coalition hunt for these remaining terrorists and for bin Laden continues. (A brief discussion of military action in Afghanistan, including Marine Corps participation, is in **Fighting Terrorism: Operation Enduring Freedom** and **Brief Military Chronology,** both later in this handbook.)

Planning for a New Afghan Government

[] In late 2001, important leaders of the anti-Taliban groups in the country and leaders from the Afghan diaspora [Afghanis living in exile abroad] met in Bonn, Germany.

[] The groups signed onto the Bonn Agreement, a plan to create a new government in Afghanistan.

[] A judicial commission was also established to rebuild the justice system according to Islamic principles, international standards, the rule of law and Afghan legal traditions.

New Government, Old Problems

Afghan Interim Authority

The beginning of the new Afghan government was the Afghan Interim Authority (AIA), which was established, with Hamid Karzai as its chairman on December 22, 2001.

International Pledges

In January 2002, to support the new Afghanistan government, sixty countries and international financial organizations present at the Tokyo Donors Conference pledged donations totaling $4.5 billion through 2006 for the reconstruction of the country.

TISA

The Afghan Interim Authority (AIA) held a nationwide grand assembly, called the Loya Jirga, in June 2002. Karzai was elected president of the Transitional Islamic State of Afghanistan (TISA) for a two-year term by the Loya Jirga. The TISA has:

- 18 months to prepare for and hold a Loya Jirga [grand assembly] to adopt a constitution for Afghanistan.
- 24 months in which to hold national elections.

In December 2002, the TISA marked the one-year anniversary of the downfall of the Taliban regime.

hllp://www.nulshellnotes.com/afghanistan lext.htm has a good update on Afghanistan, which is succinct and accurate.

Continued on next page

New Government, Old Problems, continued

Continued Factionalism

Factionalism is still a critical issue in Afghanistan even though there is one transitional government. The split between the ethnically dominant Pashtuns, some of whom were Taliban, and the "minority" tribes of northern Afghanistan, such as the Tajiks and the Uzbeks is a hindrance to real unit}'.

[] Besides this "tribal" rivalry, foreign involvement by groups vying for oil reserves complicates the problem of unifying the country.

[] Other countries still fear the spread of residual fundamentalism left from the Taliban and object to possible civil rights abuses in present-day Afghanistan.

[] The opium trade in Afghanistan also contributes to the factional problem.

Political Parties

Afghan political parties are not stable either. Karzai's government is a coalition of leaders from all political and ethnic groups. Various leaders in the country are planning to create new parties. Some of these political factions in the current government have formed new groups and parties, intending to participate in the 2004 elections.

Fighting Terrorism: Operation *Enduring Freedom*

Definition

Operation *Enduring Freedom* is the name given to US combat operations against terrorism in Afghanistan. *Enduring Freedom* began on October 7, 2001.

Purposes

The primary purpose for Operation *Enduring Freedom* is to seek out Osama bin Laden, who had been given sanctuary by the Taliban.

The Taliban who had controlled Afghanistan had allowed *Al Qaida* (Al Qaeda), the key terrorist group lead by Osama bin Laden, to establish training and base camps in Afghanistan. Thus another important reason for Operation *Enduring Freedom* is to root out and destroy these training and base camps and bring in-country members of Al Qaida (Al Qaeda) to justice.

A third purpose is to seek and destroy Taliban and anti-government forces operating in Afghanistan. As mentioned earlier, many Taliban members dispersed to mountain retreats when the coalition invaded Afghanistan. These residual fundamentalist militants are the focus of coalition search and destroy missions.

Another important reason is to ensure the stability that is necessary to reconstruct Afghanistan politically, economically, and socially. Without a stable environment, any new Afghan government will have a very difficult time uniting the country under one central government authority.

Continued on next page

Key Elements

Sea based: three strike aircraft carriers, each with various fighter squadrons aboard; the USS *Kitty Hawk,* which transported Army Special Operations helicopters; two amphibious groups built around the 15' and 26' MEUs, both of which later became the main US ground force in southern Afghanistan.

Diego Garcia base, which essentially housed B-1 and B-52 bombers.

Land basing in neighboring countries: Uzbekistan allowed US helicopters to operate out of Khanad; a few based in Tajikistan; Pakistan offered bases covertly out of Pasni and Jacobabad. These bases enabled the special forces assault on 17 October and the Marines' forward operation base (FOB) Rhino.

Precision Air Attack: Joint direct attack munition (JDAM) which Hew to preset coordinates based on satellite information (GPS/INS). Wind-corrected munitions (bomblet) dispenser (WCMD) first used in Afghanistan; laser-guided bombs. Electronic sensor surveillance used unmanned air vehicles.

A big plus was the fact that heavy bombers could engage multiple targets with JDAMs.

NATO Involvement

On September 12, 2001, the North Atlantic Treaty Organization (NATO) met in emergency session and, for the first time in its history, invoked Article 5 of the Washington Treaty. This article states that an attack against one NATO country is an attack against all member countries. As a result:

[] All 19 NATO Allies and the 9 countries applying to join NATO have provided blanket overflight rights, access to ports and bases, refueling assistance, and increased law-enforcement cooperation.

[] NATO Airborne Warning and Control System (AWACS) crews logged over 3,000 hours patrolling and protecting American skies while U.S. planes flew missions in Afghanistan.

[] Sixteen NATO allies now support Operation *Enduring Freedom* in Afghanistan and the global campaign against terrorism.

[] Fourteen allies have deployed forces in the region.

[] Nine allies are participating in combat operations in Afghanistan.

Continued on next page

NATO Involvement, continued

[] Allies and other partner countries have deployed nearly 4,000 troops to Afghanistan and have also provided 95% of the international security assistance force (ISAF), led first by the United Kingdom and now by Turkey.

[] As of August 2003, NATO took control of all security in Kabul. This is NATO's first operational commitment outside Europe.

See more details at: http://www.globalspecops.com/nalo.html

Results

Twenty-seven days after the 9/11 terrorists attacks on the World Trade Center and the Pentagon, the US joint team invaded Afghanistan and two months later freed Afghanistan from Taliban domination, together with US allies.

This was accomplished without:
[] Any US bases in-country
[] A major plan for removing the Taliban from power.

http://www.lejeune.usmc.mil/2dmardiv/36/ has detailed reports and lessons-learned from Marne Corps experience in Afghanistan. Click on the link above; click on operations, then on professional forum. Then click on any of the reports individually.

Continued on next page

Fighting Terrorism: Operation *Enduring Freedom*, Continued

Key Accomplishments

- [] The deepest amphibious operation in Marine Coips history - over 400 miles. Task Force 58 conducted operations in Afghanistan far from the sea. The operations showed that Marines could operate expeditionary forces at extended distances from a sea base.

- [] Marines and soldiers fought battles at extremely high elevations, over 10,000 feet above sea level.

- [] *Enduring Freedom* was the first time C-17's were employed in a medium threat environment to air drop supplies.

- [] For the first time, headquarters and the aircrew viewed a picture taken from an unmanned vehicle simultaneously over the target.

- [] Operation *Enduring Freedom* included the US Air Force's longest combat sortie on record - 44 hours. Over all, naval aircraft flew about 75% of all sorties and dropped about 33% of the bombs. The Air Force flew about 25% of the combat sorties and dropped about 75% of the bomb tonnage, including the "Daisy Cutter."

Brief Military Chronology

Introduction

This section is an abbreviated discussion of military action during Operation *Enduring Freedom*. The topics include the following:

Early Days

On October 7-8, 2001, The United States and Great Britain began strikes against Taliban air defenses, fuel sites, al Qaida (Al Qaeda) terrorist training camps and headquarters and military installations.

[] The main thrust of air assaults was the Taliban regime in Afghanistan: Concurrently crippling Taliban movement, the allies supported the Northern Alliance

[] Such precision bombing only stiffened the Taliban who did not hand over Osama bin Laden.

These targeted actions were designed to disrupt the use of Afghanistan as a terrorist base of operations and to attack the military capability of the Taliban regime. At the same time the U.S. and its allies began to drop food, medicine, and supplies to the population of Afghanistan.

Continued on next page

Brief Military Chronology, continued

War in the North

Two issues moved the war effort forward in northern Afghanistan: increased bombing by the allies and the commitment of US Special Operations Forces (SOF). This same kind of commitment was to be important in the success of the Marines later in their search for residual Taliban and Al Qaida forces.

[] On November 9, Mazar-i-Sharif fell to the Northern Alliance, effectively opening a land route directly to Uzbekistan and blocking any northward movement of the Taliban from Kabul to Uzbekistan. Capturing this city also offered coalition forces a major airfield in-country.

[] The momentum of the Northern Alliance supported by the allies rolled on until the Taliban abandoned Kabul on 12 November.

[] The last Taliban stronghold, Konduz, surrendered on 26 November.

War in the South - FOB Rhino

Background

The Taliban had a stronger hold on southern Afghanistan. It was more difficult in the south to muster the local population to oppose them than in the north. The Taliban also had better security forces there.

MC Mission

The plan called for TF58 to seize and hold Rhino, while conducting raids from it. As General James N. Mattis, commander of TF58, expressed it:

"...to create chaos, denying the enemy their sense of security."

This coupled with guidance from Vice Admiral Moore, Commander NAVCENT/ MARCENT: The Marine raids would defeat the Taliban and Al Qaida forces quickly and decisively became the Marines' mission.

Elements

US Army Special Forces: Generally, these forces could initially seize an airfield or installation, but did not have the manpower to hold it. This is where Marine Corps forces came into play.

Marine Corps: Task Force (TF) 58 consisting of the
[] 15th Marine Expeditionary Unit (SOC) [MEU (SOC)]
[] 26th MEU (SOC).

Location

An airstrip that the Special Forces had raided, called FOB Rhino.

Continued on next page

Establishing the FOB

Marines seized FOB Rhino on November 25, 2001.

[] They linked up with Navy SEALs who had earlier arrived there for target surveillance and reconnaissance.

[] Once the airfield was declared KC-130 capable, additional Marines were landed. The MEU's attack helicopters provided close air support.

TF 58: Interdiction of Rt. 1

This operation began on December 4, 2001, and included an approximately 85-mile, 19-hour motorized movement across the desert. The raid resulted in the:

[] Capture of a Taliban convoy.
[] Detection of a P-3.
[] Killing of 120 enemy personnel.

Capturing Kandahar

Background

Kandahar was a Taliban stronghold, the center of their power. The Taliban refused to surrender the city to Afghan Southern Alliance forces, and Al Qaida began to slaughter local Afghanis who were surrendering. Finally, on December 6, 2001, the city fell.

MC Action at Kandahar Airport

TF58 linked up with the Southern Alliance and U.S. Special Forces on December 13, 2001.

[] By December 14, TF58 had established a second FOB at the Kandahar airport.
[] Marines repaired the airstrip so that by December 18 the airport was considered safe for C-17s.

Further MC Operations

Looking for intelligence materials:

[] In villages west of FOB Rhino
Near Kandahar

War in the South: Khowst-Gardez Mission

Background

Although Kandahar had fallen to the alliance and coalition allies, strong pockets of resistance still remained. The local alliance leader had merely raised the alliance flag; his troops had not actually taken on the Taliban.

TF58's Mission

TF58 was tasked to seek out and destroy these pockets. Special Forces would identify the targets, and Marines would "hammer" them. Because the distance from Kandahar to the battle space was about 250 miles, a local airfield, about an hour from the battlefield, was used for close air support.

Disestablishing TFS8

During January 2002, TF58 was relieved in place by troops of the Army's 101st Airborne Division; however, when the Taliban attacked the Kabul airport, the Marines beat them back. TF58 was disestablished at the end of February 2002.

Tora Bora and Operation *Anaconda*

Al Qaida (Al Qaeda): Two Options for Escape

Al Qaida, including Osama bin Laden, and surviving Taliban leaders and forces had had two possibilities for escape from invading coalition forces:

[] They could leave Afghanistan by escaping to Pakistan or through Pakistan to countries that harbored groups sympathetic to the Al Qaida cause, such as Somalia, Indonesia, or the Philippine Islands.

[] The second possibility was to go into hiding in the mountain caves and retreats in the Tora Bora Mountains in eastern Afghanistan, near the Pakistani border.

Tora Bora

The assault on Al Qaida (Al Qaeda) in Tora Bora was conducted by Afghan troops strongly supported by U.S. firepower. Few U. S. troops were directly involved in the assaults. Afghan troops took the caves, but many of the key Taliban and Al Qaida figures there escaped.

About 300 bodies of bin Laden fighters were counted; other bodies were probably crushed beneath collapsed caves and bunkers. 150 more enemy personnel were caught trying to escape across the border into Pakistan.

Tora Bora taught Al Qaida forces that U. S. detection systems could intercept messages from radios and cell phones. Al Qaida would attempt to profit from this lesson-learned in Operation *Anaconda.*

Operation *Anaconda*

Operation *Anaconda* (in Gardez and Shah-e-Kot Valley) was mainly a U. S. forces operation. British, Canadians, and Australians also took part in the action. There were about 1,000 Afghan troops, 1,000 U. S. and some additional coalition members.

The purpose of the operation was to destroy what was considered the last pocket of Al Qaida troops, thus preventing them from launching an offense the following spring.

Continued on next page

Tora Bora and Operation Anaconda, Continued

Tactics

Limited attacks on Al Qaida (Al Qaeda) troops to force them to reveal their positions so that air strikes could be called in.

[] Apparently, Al Qaida troops attacked whenever they saw U.S. troops, regardless of the conditions of the battlefield, e.g., they would attack uphill and into fire.

[] Al Qaida also had little understanding of the sensors and sophisticated weapons they faced.

Cave Raiding: Results

Although there were some problems, in general, this "cave raiding" was worthwhile:

[] Documents and computers, with clues to future Al Qaida (Al Qaeda) operations as well as lists of Al Qaida personnel and financial accounts of the group were recovered.

[] Information about the Al Qaida's interest in weapons of mass destruction was also forthcoming.

Protecting the interim Afghan Government

In December 2001, The United Nations assembled the International Security Force (ISAF) to stabilized Kabul, the capital of Afghanistan. The bulk of the ISAF came from Muslim troops from Turkey and Jordan, with small contingents from the Canada, Western Europe and Australia and New Zealand.

Note: Much of this brief military chronology came from chapter 9 "Striking Back" of Norman Friedman's Terrorism, *Afghanistan, and America's NEW WAY OF WAR.*

Current Challenges

Introduction

Rather than try to explain the current challenges of deploying to Afghanistan, a number of "case studies" are detailed below. You should be able to get some idea of the "working" environment in country that Marines have already encountered and that you may have to face. The topics in **Current Challenges** are shown in the table below:

Local Problem: Low Intensity Civil War

"Karim Khan stands disconsolately outside the local government headquarters in the remote village of Tuksar [in Afghanistan! • *He used to run the neighbouring village, but was bundled out by a rival militia one night recently, leaving his wife and family behind as virtual prisoners."*

[] The incident is not isolated. It is not so uncommon throughout northern Afghanistan, where central government control is not as strong as it is near the Afghan capital Kabul.

[] This has led a number of experts to consider some aspects of the combat situation in Afghanistan today as low-level civil war in which local militias use the autumn, the country's traditional fighting season, to change the map of power.

Continued on next page

Feuding Warlords

Government curbs northern warlords

Feuding commanders' armies to merge as Kabul shows resolve.
By Ahmad Nahim Qadiri in Mazar-e-Sharif (ARR No. 79, 29-Oct-03J

"Afghanistan's transitional government has taken decisive action to improve security in the north of the country, dismissing key officials in Balkh province and ordering two feuding military forces merge. In a major shake-up announced by interior minister Ali Ahmad Jalali on October 26, the governor and deputy governor of Balkh province were replaced and the chief of police in the administrative centre Mazar-e-Sharif was dismissed.

"Jalali also announced that the two army corps run by rival commanders General Abdul Rashid Dostum and General Atta Mohammad are to be merged into a single force and integrated into the new Afghan national army. Disarmament of heavy and light weapons is scheduled to begin in a few weeks in Mazar-e-Sharif, supervised by the national defence ministry, the United Nations Mission in Afghanistan, UNAMA, and the British-led provincial reconstruction team, PRT.

"Analysts in Mazar-e-Sharif say the changes are designed to strengthen the hand of President Hamed Karzai's government in the north of Afghanistan, ahead of crucial decisions on a new constitution and elections planned for next year. They say the decisions are important for symbolic well as practical reasons. ... The changes come amid heightening tension in Mazar-e-Sharif, where there were fears that a ceasefire agreement between the forces of Dostum and Atta Mohammad could collapse.

"Dostum, who leads the Junbesh-e-Milli movement and has been a key player in the north for more than a decade, has been battling against Atta Mohammad for control of Mazar-e-Sharif and surrounding territory since the fall of the Taleban regime nearly two years ago.

'The two men control large forces — designated the Eighth and Seventh Corps, respectively - which formally come under Defence Minister Mohammad Qasim Fahim, who like Alia Mohammad is part of the powerful Jamiat-e-Islami faction. The turf war between the two groups has frustrated attempts by the transitional government to extend its authority to the northern provinces."

As you can see, this article points up a number of issues that in-country deployed forces may have to deal with:

[] Indigenous forces that are not united
[] Tension between the central government and locally based leaders
[] Local vice the national view on demobilization
[] Turf-war claimants

From:
http://ww.reUefweb.int%/rw^^
256dcl"0012()l"a8?Open] document

Continued on next page

Introduction

As a member of 3d Bn, 6th Marines, L Co., 1st Plt., 26 MEU (SOC), these are the experiences of an infantry platoon sergeant who served in Afghanistan:

My platoon was the security element for the company. We arrived in Afghanistan on 19 Dec 2001 and stayed until 5 Feb 2002.

FOB Rhino: Insertion

- We flew from Pasni, Pakistan, into FOB Rhino on 19 Dec 2001. We went from 90-degree heat to 40-degree chill in 3 hours. What an adjustment! After the first couple nights we became accustomed to the drastic temperature change.
- The biggest challenge was the feeling that we were isolated. Since 1/1 was securing the outer perimeter we only had to man the interior posts. After 2 days we were flown by C-130 to Kandahar.

Kandahar

Purpose

Our purpose at Kandahar was to augment security of the airfield:

- We didn't have much interaction with the civilians.
- The civilians we did meet seemed very pleasant and not hostile in any way.

Our translator told us the people were very glad to have security and stability in their lives for once.

Climate

The temperature in January ranged from 50° F. in the day to 15 °F. at night. You can expect a 20-30 degree temperature change when the sun sets.

It was dry except for two rainy days. When it rains, the sand and hard packed dirt become ankle-deep mud. The mud is slippery and it stinks!

Kandahar (Continued)

Keeping Warm

Marines will find ways to stay warm no matter what. Here are a few ways we kept warm.

[] While on watch Marines would lay their poncho liners over their laps.
[] Hot wets were prepared before nightfall and served out to the Marines.

These little things plus the gear we brought with us made our nights passable.

Keeping Clean

Since we didn't have showers, and water was at a premium, we had to take canteen cup baths, better known as birdie baths.

[] We would heal up a canteen cup full of water and wash ourselves head to toe.

[] With that same canteen cup of water we would then wash our socks and tie them to our packs to dry in the daylight.

After a few weeks we received wash buckets to wash our clothing. Teaching Marines how to wash clothes by hand and with a washboard was unique!

Making Life Bearable

We had packed enough tobacco products to last 30 days. Since the army took nearly a month longer to get there, we made arrangements with some air force personnel to bring some tobacco in to us once in a while.

Mail didn't arrive in the country until the week before we departed. Priorities were water, food, equipment, ammo, fuel, and then mail. However, the MEU did set up three to four phones for us to use. We would have to wait for hours to make a call, but it was worth it.

Coffee was made by pulling a clean sock tip around the top of a canteen cup; put the coffee grounds on the sock, then pour hot water over the grounds. Coffee made a huge difference!

Continued on next page

Kandahar (Continued)

Dangers

While in Kandahar there were a few things we, Marines, always kept in the forefront of our minds: mines, mines, and more mines.

[] Not only did we have to watch for Soviet mines, we had to be careful of our own cluster bomblels. Wherever we saw some in the area, we had our Marines take a look at Ihem so Ihey would know what they looked like.

[] Unexploded ordinance was everywhere. EOD was busy with the bigger stuff. All our Marines could do was mark near where the ordinance were and avoid the area.

Important!

Some precautions that paid off:

[] Gear inspections and gear accountability (No one lost any gear!)
[] Uniformity within the company
[] Entrenching was continuous. We made bunkers not just fighting positions.
[] Weapons training (not one ND while we were there!)
[] Training the NCOs (Never did I catch a Marine breaking light discipline or falling asleep on watch) The NCOs ensured their Marines were alert and prepared!

Raids near Khowst

Purpose

Our purpose near Khowst was to conduct a raid on cave complexes located near there. The raid was supposed to last under 10 hours from insertion to extraction; however, it lasted 10 days.

[] In those 10 days we gathered much of our intelligence information.
[] We also proved that Marine units could and did survive, fight, and prevail without resupply.

Climate

Khowst is located in northeast Afghanistan, near the border with Pakistan. The climate in the mountains at night was extremely cold, in the teens to single digits. During the day the temperature was 40°F-50°F. There were trees and scrub brush in the area. It didn't rain during the time we were up there.

Continued on next page

Raids near Khowst (Continued)

Terrain

The terrain is mountainous. Physical preparation and education is needed for operating in a mountainous environment.

- [] Vehicle movement is restricted to the *Wadis* and roads,
- [] There are many old bunkers, fighting positions and caves throughout the mountains.
- [] There were trees and scrub brush in the area.

Enemy Tendencies

The enemy made bivouac sites opposite their caves. The cave locations were mostly in the *Wadis* where they were extremely difficult to see or bomb.

- [] We found equipment slashed in burial sites, buried outside villages, and just left plainly in the open.
- [] The enemy signaled at night from mountaintop to mountaintop with white light.
- [] There was also evidence that they had cell phone/satellite phone capability.

As soon as we landed, we found evidence thai the enemy had hastily withdrawn into Pakistan. After a few days we had a helicopter arrive to take detainees to Kandahar. Some started making their way back over the mountains.

Interacting with the Local Population

Some of the people we took as detainees were absolutely scared of us. Based on our physical size and appearance, they would become almost hysterical, but with a little show of kindness on our part, they willingly told us many tidbits of information: All we had to do was give them a Koran and let them know when it was prayer time. All of a sudden we discovered that one of them could speak English.

- [] These people knew which group controlled what area; they would claim allegiance to both so thai they could have unfettered access to both areas.

- [] The people seemed to side with whoever was more powerful.

- [] They have little national pride; they did have pride in their tribes!

In a vehicle we captured going down the ***Wadi*** we found a Taliban flag and a Northern Alliance flag. In the same vehicle we also found a GPS.

Continued on next page

Raids near Khowst (Continued)

Living off the Land

Because we had no re-supply, we had to live off the land (a 10 hour-raid means you carry very little gear).

[] We seized an abandoned terror training facility and found livestock and grain.

[] We soon began making corn bread, pop corn, and other types of food. After a couple days we had chicken, lamb, and steak.

[] We found water in cisterns and in natural springs in the *Wadi.* We purified the water in 5-gallon jugs.

Small Unit Operations

In the mountains we had great success, operating at squad size and less. There were always 2 patrols out day and night. They would range 3-4 km and stay at altitude. This helped us gain vital information of enemy movements and discover weapons caches.

We constantly used binoculars and NVGs with 3x Mag for observation. This timely warning and information meant we could capture many people who may have otherwise become combatants.

http://www.lejeune.usmc.mil/2dmardiv/36/ has detailed reports and lessons-learned from Maine Corps experience in Afghanistan. Click on the link above; click on operations, then on professional forum. Then click on any of the reports individually.

Continued on next page

Current Challenges: Joint Operations, Continued

Broad Tactical Environment

By Jim Garamone

American Forces Press Service
WASHINGTON, June 4, 2003 - U.S. and Italian forces launched an offensive June 2 against anti-coalition forces in the Shahi Kovvt section of the Paktia province in Afghanistan, Combined Joint Task Force 180 officials said today.

A brigade-sized unit launched Operation Dragon Fury to prevent the re-emergence of terrorism, to deny sanctuary to terrorists and to protect nongovernmental organizations and coalition forces from enemy attack.

Army Col. Rod Davis, spokesman for the command, said the ground assault section set up blocking positions and air assault troops drove anti-coalition forces into the trap. A total of 21 men were taken under control by coalition forces.

The U.S. forces came out of Kandahar, the Italian forces from Khowst. Paktia province is located in eastern Afghanistan near the border with Pakistan. The province was the scene of tough fighting in 2001 and was raided again last year.

Roughly 20 helicopters launched the air assault. Close-air support was standing by in case the coalition troops needed the assistance, Davis said. He said the Italian forces searched more than 300 vehicles and inspected more than 800 people.

There were no coalition casualties.

American forces came from the 82nd Airborne units in the area. Special operations personnel also took part.

From: http://www.globalspecops.com/oef2a.html This article shows the "larger environment" in which Marines like those in the earlier case study might be involved in.

Rebuilding Afghanistan

Armed Forces

The United States is helping the Afghan government to build a national army. In 2002, US soldiers helped train 1,600 Afghan soldiers for this new army.

http://www.elobalspecops.com/rebuilding.html is good site that summarizes US efforts to rebuild Afghanistan.

Assistance to Women

Although the extreme persecution of women inflicted by the Taliban has ceased, women still have a difficult lime in presenl-day Afghanistan.

[] The U.S. has helped the Afghans to establish women's resource centers.

[] Over one million dollars has been used for training women in business and for educating girls.

Transportation Infrastructure

The United States has committed $800 million to reconstruct the main transportation route between Kabul, Kandahar, and Herat. The project, underway since 2002, is scheduled for completion by 2004.

The United States is helping the Afghan government to repair and reconstruct bridges and 4,000 km of secondary and tertiary roads. The United States is also helping to winterize a major tunnel that is a key link to northern Afghanistan as well as participating in revitalizing over 6,000 water wells, canals, dams, and water systems in the country.

Child Welfare

[] The United States has funded an immunization program for Afghan children. Since April 2002, 4.3 million Afghan children have been immunized against measles.

[] America's Fund for Afghan Children has raised over $11.4 million to help Afghan children. Moneys from this fund have been used to purchase school supplies and books as well as teacher kits. Funds have been used to build playgrounds. Winter relief and health kits for children have also been supplied from this fund.

Continued on next page

Refugee Welfare

More than two million refugees and displaced persons have returned home to Afghanistan since the overthrow of the Taliban.

- The United States has donated funds and assistance to help resettle these refugees and to provide them with food and basic needs, as well as shelter kits to the neediest.

- An Afghan Conservation Corps has been established, giving Afghan refugees who have returned to the country and demobilized fighters employment and job training.

Health

- The central disease surveillance system for the eradication of polio was re-established, and a national curriculum for midwives was revised.

- Health facilities, including ov er 28 clinics and hospitals, have been rebuilt and refurbished. A national plan has been drawn up to determine the location and type of construction for an additional 600 health care facilities.

Economic and Agricultural Activities

The Afghan government has an ambitious business enterprise policy. The aim is to restore full central government authority over the nation's currency and fiscal system. The United Slates is assisting the Afghans in developing banking, trade, and fiscal and business reforms. More privatization of stale owned banks and public sector organizations is also sought.

Seventy percent of the Afghan population is engaged in agriculture. The United States has funded assistance to Afghan farmers aimed at improving crop production. In the spring of 2002, the United States donated 7,000 tons of seed, fertilizer and technical support to over 40,000 farmers. During 2002, crop production increased 80% over 2001 production.

http://vvvvvv.developmentgateway.org/node/134111/ is an excellent site on reconstruction in Afghanistan, showing what various coalition members are doing to help the country.

Islam: Basic Ideas

Introduction

You have probably heard the name *Islam* associated with Afghanistan. Most people in Afghanistan and in surrounding countries believe in Islam. The Taliban and Osama bin Laden believe in a radical form of Islam. In this section, you will learn about Islam, its main ideas, customs, and practices.

(http://thetruereligion.org/intro.htm - - a short introduction to Islam)

Definition of Islam

Islam means submission to the will of God in the Arabic language. The word *Islam* has the same root as the Arabic word *salaam,* which means peace. Muslims around the world often greet each other with the phrase "Salaam" in much the same manner as Jews say, "Shalom," as a greeting. Both salaam and shalom come from the same common language root, the three-letter root -- *s-l-m.*

(http://media.isnet.org/off/Islam/basics/index.html) - discussion of various "doctrinal" aspects of Islam by noted theologians.

Allah

Islam is not named after a person like Christianity, which was named after Jesus Christ, Buddhism after Buddha, and Confucianism after Confucius. The central focus of Islam is always God.

The basic tenet of Islam is that you must submit to Allah (the Arabic word for God) and live according to His divinely inspired law.

The Precept

- The key truth that Allah has revealed to mankind is that the only divine and worshipful "being" is Allah, the almighty God; thus all human beings should submit to and worship Allah.

- Allah has 99 names, among them, The Gracious, The Merciful, The Beneficent, The Creator, The All-Knowing, The All-Wise, The Lord of the Universe, The First, The Last, and others.

No matter what "sect" of Islam a person belongs to, he believes in this important religious belief that is the common thread of Islam.

Arabs and Islam

Muslims (Moslems)

Now that you know that the religion is called Islam and the name of the God of Islam is Allah, what does the word *Muslim* mean? The word has the same s-l-m root as the words *Islam* and *salaam.* A Muslim is a person who submits to the will of Allah, regardless of race, nationality, or ethnic background. A Muslim has to submit completely and obediently to Allah and live according to Allah's message.

[] Muslims worship Allah alone and must not worship any person, place or thing other than Allah. They believe that Allah is the God for the Christians, the Jews, the Muslims, the Buddhists, the Hindus, and even atheists.

[] Orally repeating the basic belief of a Muslim found in the "motto" of Islam: *"Laa Elaaha illa/laah"* which means, "There is no god but Allah," is the way a Muslim professes that he belongs to Islam.

Importance of Arabic in Islam

The "motto" above, *"Laa Elaaha illatlaah"* is in Arabic. Even though the people of Afghanistan speak a number of different non-Arabic languages, the language of Islam is Arabic, and this is the language used in "religious services." (Historically, Islam began among the Arabs of what is now Saudi Arabia.)

[] The prayers, sayings, and many blessings of Islam are all spoken in Arabic, just as the holy book of Islam, the Qur'an (Koran) is written in Arabic.

[] This common religious language of Islam is one of the ties that bind all Muslims together.

Continued on next page

Arabs and Islam, Continued

Arabs and Non-Arab Muslims

Because of the importance of the Arabic language in Islam, many people assume that most Muslims are Arabs. This is not the case: About 80% of all Muslims are **not** Arabs.

[] In fact there are more Muslims in Indonesia than in the Arab Middle East.

[] Besides Indonesia, Muslims make up the majority in such non-Arabic countries as Turkey, Pakistan, Uzbekistan and other central Asian countries of the former Soviet Union.

[] There is a large minority in China too.

[] In Europe, Albania is Muslim. Bulgaria, Bosnia, and Macedonia have large Muslim populations. Elsewhere in Europe, there are immigrant communities of Muslims from Africa, Turkey, and Asia in France, Britain, and Germany.

[] In the Americas Muslims have increased in recent years, both from conversions and immigration; 20% of the population of Suriname in South America is Muslim. Today there are about five million Muslims in America.

Muhammad (Mohammed)

Importance

Muhammad is ... the Messenger of God
and the last of the prophets ... (Qur'an, 33:40)

This one phrase from the Qur'an shows Muhammad's place in Islam and in world religion. Muslims believe that Muhammad is the prophet of Islam. His message is Islam. The revelation of Islam that Muhammad received is the *Qur'an* (Koran).

Muhammad and Other Religions

In terms of other world religions, Muhammad is the last prophet of God to mankind.

[] He is the last messenger of God.
[] His message applies to all mankind, regardless of their beliefs.
[] Muhammad is the successor to Moses, Jacob, Isaac, Abraham, and Jesus.

Muhammad is the prophet among all previous prophets and messengers.

[] He reinterpreted and corrected the message of Allah and gave mankind the true word from Allah. In a sense, Muslims consider that he "cleansed" the message.

[] In spite of his importance in Islam and for the world, Muslims do not worship him or ascribe any divine qualities to the prophet.

Muhammad, His Calling

Muhammad was born in Mecca (Makkah) in what is now Saudi Arabia, in the year 570 CE. His father died before he was born. Soon after his birth, his mother also died, so Muhammad was raised by his uncle.

Because of his trustworthiness and sincerity he was often asked to arbitrate disputes. He has been described as calm and meditative. Muhammad was very religious. It was his habil to meditate in a cave near Mecca.

At the age of 40, during one of his retreats, Muhammad received his first revelation from Allah through the Angel Gabriel. This revelation, which he continued to receive for twenty-three years, is known as the *Qur'an* (Koran).

Continued on next page

Muhammad (Mohammed), Continued

Spreading Islam

As Muhammad began to preach the truth that Allah had revealed to him, he and his small group of followers suffered persecution.

[] This became so strong that in the year 622 CE Muhammad and his followers emigrated from Mecca.

[] This event, called the *Hijra* ('migration'), when they traveled about 260 miles to the north to the city of Medina, marks the beginning of the Muslim calendar.

[] After several years, Muhammad and his followers returned to Mecca, where they made peace with their enemies and established Islam.

[] Before Muhammad died at the age of 63, most of Arabia was Muslim, and within a century of his death, Islam had spread to Spain in the West and as far east as China.

Five Pillars of Islam

Introduction

The five tenets are the foundation for all Islamic belief. They are the cornerstones of Islam. (http://www.usc.edu/dept/MSA/fundamentals/pillars/ - an introduction to the five pillars.)

Commitment

Called the Ash-Shahaadah, this is the "credo" of Islam:

"I bear witness thai there is none worthy of worship except Allah and I bear witness that Muhammad is His servant and messenger."

All Muslims subscribe to and repeat this belief.

Prayer

Called As-Salaat in Arabic, this is the requirement that adult Muslims pray at five specific times during a twenty-four hour day to Allah.

[] This is a ritual prayer requiring Muslims to say specific prayers and to coordinate hand and arm movements, eventually culminating in complete prostration or bowing down with the hands on the floor to show complete submission to Allah.

[] While performing As-Salaat, Muslims face in the direction of Mecca.

Alms

Known as the Az-Zakaat in Arabic, this means that Muslims must contribute to the support of those less fortunate Muslims—support of the poor. Muslims are encouraged to pay their Zakaat during the holy month of Ramadan.

Continued on next page

Five Pillars of Islam, continued

Fasting

In Arabic this is called As-Sawam. It means that during the month of Ramadan, adult Muslims who are in good health

[] Cannot eat, drink, or smoke from sunrise until sunset every day of this holy month.

[] Must abstain from sexual relations.

The intention of fasting is to worship Allah. Ramadan is a sacred month because it commemorates the time when Allah first revealed the Qur'an to Muhammad. The time for Ramadan changes each year because it is based on a lunar calendar. Ramadan comes 11 days earlier each year.

Pilgrimage

Called Al- Hajj (Hadsh), this pilgrimage requires a Muslim in good health to travel to Mecca in Saudi Arabia at least once in a lifetime.

[] Specific preparation is necessary for the Hajj; for example there are rules about shaving and cutting hair and what to wear.

[] One of the high points of the Hajj is the *umrah,* which means performing a ritual walk seven times around the sacred stone of Islam called the Kab'ah.

[] The Hajj commemorates the sacrifices and faith of Abraham, his second wife, Hagar, and their son, Ishmael.

According to the Council on Islamic Education, it is the largest, regularly scheduled international gathering on Earth.

Al Qur'an (Koran) and Sunnah

Definition

Al Qur'an (the Qur'an) is the revealed word of Islam, the holy book of the Muslims.

[] Muhammad received the message in bits and pieces from the Angel Gabriel during a period of 23 years.
[] The word *Al Qur'an* means the recitation. During Muhammad's lifetime it was recited publicly.
[] The original language of the Qur'an is Arabic.

It is important to remember that Muslims believe the Qur'an is **the** word of Allah, untouched and uninterpreted by human beings. As Allah's final message to mankind, it is the holy book that supercedes all others: the *Old Testament,* the *Gospels,* etc.

Form and Content

The Qur'an is a collection of surahs or chapters, many of which you might call verses.

[] Muhammad received the surahs from Allah. Scribes selected by Muhammad usually wrote down these verses. Sometimes they wrote on wood, trees, parchment, and even stones. Many followers also memorized the Qur' an by heart.

[] In later years, the Qur'an was recopied and refined; accents and markings for reading were added.

Interpretation

[] The Sunnah is the spoken word and acts of Muhammad. The Sunnah, Muhammad's actual words and deeds, uses the life of Muhammad to explain and expand on the verses and teachings of the Qur'an so that Islam can eventually become a world religion.

[] A second form of interpretation is called hadith. This is narration about the life of the prophet or what he approved - as opposed to his life itself, which is the Sunnah.

The Qur'an, the Sunnah, and the ahadith (plural of hadith) are the basic body of Islamic religious doctrine for all Muslims. (http://www.afghan-network.net/Islam/) (http://www.usc.edu/dept/MSA/)

Jihad (Djihad)

Definition

The word *jihad* means straggle in Arabic. This means a straggle between the forces of Allah (good) and the forces of evil —an eternal struggle.

You have probably heard the word *jihad* in reference to Osama bin Laden and the Taliban. They consider the terrorism that they perpetuate a jihad. Actually, the word *jihad* has a couple of interpretations.

Self-defense

Islam allows fighting in self-defense and in defense of religion. There are rules of combat including prohibitions against harming civilians and against destroying crops, trees and livestock.

If good Muslims do not risk their lives in the cause of Islam, injustice would triumph. The Qur'an says:

> " *Fight in the cause of God against those who fight you, but do not transgress limits. God does not love transgressors.*" (2.190)

War is the last resort and is subject to the precepts of Islamic sacred law.

Inner Struggle

The second and more personal meaning of jihad is an inner straggle. Each of us, according to Islam, is constantly fighting a "war" against self-centered desire, and egotism, with the final goal of attaining inner peace.

Islamic Schools (Sects)

Religious vice Secular Law: Western View

The United States and most other Western countries make a clear distinction between the rule of law and religion.

[] Religion is a separate personal issue that does **not** determine law.

[] A legal case may make use of religion, but religion is not the determining factor in most legal cases.

[] Simply put, a religious denomination, such as Methodism, Catholicism, or Judaism, doesn't have the power to rule on a case of adultery in a civil court or on a case of fraud in a criminal court.

The United States has a secular system of law; religion plays no direct part in such a system.

Islamic View

If you look at Islam, the perspective is different. Islamic law consists of guidelines and rules that determine all aspects of a Muslim's life from how to perform ritual prayer to conducting business transactions.

[] It also includes crimes and the appropriate punishments for each.

[] These Islamic laws are based primarily on the Qur'an and are called the Sharia.

[] Muslims must apply these precepts handed down to Muhammad in the sixth century to present-day situations.

Interpretation and Re-interpretation

This kind of religious legal system, particularly how to interpret the Qur'an and the ahadith, has resulted in a certain degree of divisiveness in Islam, just as varied interpretations of Christian doctrine have caused Christianity to split into various denominations. The two main sects of Islam, which are the schools of Islam, are the

[] Sunni
[] Shi'ites (Shi'ia)

Continued on next page

Islamic Schools (Sects), Continued

Shi'ites (Shi'ia)

The basic divisiveness in Islam is based on an historical incident:

[] After Muhammad died, there was confusion about who should succeed him as the leader of Islam. One group felt that the successor was Imam Ali who had been appointed by Muhammad according to Allah's decree. Ali was a relative of Muhammad and the first to accept Islam.

[] This group, mostly from the household of the prophet, was a minority and became known as the Shi'ites. They also believe that the leader of Islam must be endowed with grace and benevolence and should be infallible.

Sunnis

The majority group, called Sunni Muslims, believe that:

[] Muhammad did not choose a specific successor and probably assumed that after his death, Muslims would find their own leader.

[] The prophet did not tell his followers how they ought to select their leaders or what qualifications their leaders should have.

The Sunnis chose a leader, later called the caliph, from outside Muhammad's household. The Shi'ites refused to accept the Sunnis' choice and split off from this main group.

Results of the Division

Because of this succession split, variations in Islamic doctrine, law, and practice have developed between these two main groups over the centuries.

[] Sunni Muslims predominate in Saudi Arabia, Afghanistan, Pakistan, Turkey, and Indonesia.

[] Shi'ites are the majority in Iran and southern Iraq.

Sunnis make up about 83 percent of Muslims, according to the *Encyclopedia Britannica;* Shi'ites, about 16 percent; and a few other small groups, the remainder.

Osama (Usama) bin Laden and Islam

Bin Laden's Religious Background

Bin Laden's brand of Islam is based on extremely conservative Islam called Wahhabism, which was espoused by Muhammad bin Abd al-Wahhab (1703-87) He was the founder of the sect and the co-founder of Saudi Arabia. The Wahhabis and the royal Saud family have long intermarried. The Saudis have adopted many of the conservative precepts of Wahhabism.

Wahhabism: The Way of a Muslim Life

Wahhabi Islam is a total religious system. It has an answer for every question about a Muslim's life including social, legal and spiritual aspects. Wahhabi Islam is ascetic.

[] Men should wear short robes and even avoid the black cords used on turbans and headgear.
[] Mosques should have no decoration.
[] Drinking alcohol is forbidden.

Punishment is based exclusively on the Qur'an, for example:

[] The right hand should be amputated for theft.
[] Adulterers should be stoned to death.
[] Murder and sexual deviation are punishable by beheading. To this day Saudi Arabia uses these punishments, especially beheading for capital crimes.

Bin Laden and the Taliban

The kind of radically, conservative state that the Taliban had developed in Afghanistan had roots in Wahhabism, which inspires and feeds on the Taliban's fundamentalist religiosity.

[] For Islamic fundamentalists, the Taliban have created what approaches a pure, ideal society based on Islamic religion and law.
[] This society fits bin Laden's mold too.

The Taliban allowed bin Laden to remain in Afghanistan as a guest. They denied consistently that he had anything to do with the terrorist acts of 9/11 and refused to extradite him as requested by the US government. They continuously claimed that they did not know of his whereabouts.

Continued on next page

Focus on Expansion and Terrorism

[] While bin Laden follows the conservative Wahhabi tenets, what he most fervently supports is that his brand of Islamic faith must expand. (To support this kind of expansion, Saudi Arabia supported the Afghan *mujahidin* when they were fighting the Communists. So did Osama bin Laden.)
http://www.newyorker.com/FROM_THE_ARCHIVE/ARCHIVES/7010924 IV_archi\ c() - good article on bin Laden, his life and religion

[] Now bin Laden regards the struggle as a war between two civilizations: His Islam and the non-Muslim civilization, specifically the United States. Toward winning this struggle, bin Laden has committed his forces to terrorism.

http://www.brookings.Org/fp/proiects/terronsm/faqs.htm#qb4 - a set of informative questions about terrorism.

Bin Laden-sponsored Terrorism

[] Besides the infamous 9/11 terrorist attacks against the Pentagon and the twin towers, bin Laden allegedly sponsored the attacks on American embassies in Africa.

[] U.S. and Russian officials arc also concerned that he is financing Chechen rebel operations out of Dagestan, a former Soviet republic in Asia.

A primary concern is that bin Laden's forces are attempting to acquire or build chemical and/or biological weapons to use in their terrorist attacks. In the case of the Chechen rebels, he may be trying to help them acquire radiological dispersal devices.

Continued on next page

Al Qaida (Al Qaeda)

Al Qaida (Al Qaeda) is the name of the network of Islamic extremists that bin Laden has at his command to carry out his radical Islamic terrorism.

[] Al Qaida consists of a group of about 3,000 commanders. Troops of Afghanis and Pakistanis number 200,000+.

[] There are a number of terrorist centers, with numerous cells, in North America, Yemen, Saudi Arabia, Albania, Kosovo, Algeria, Chechenya, all the former Soviet republics of Central Asia, the Philippines, Egypt, Ethiopia and Somalia.

Bin Laden's "Armv"

Al Qaida is bin Laden's personal "army." Bin Laden could probably rally an army of over 100,000+ men around the world, excluding those currently "in service" in Pakistan and Afghanistan.

[] Many members of his overseas forces are ready for both sustained and one-time operations.

[] In the interim, the members of these terrorist forces return to normal "civilian" life in host countries. They are also available for terrorist operations on their own countries.

Funding for Al Qaida (Al Qaeda) Terrorism

The primary sources for funding Al Qaida-sponsored terrorism have been:

rj -pncsa|c0f drugs, primarily opium.

[] Contributions from Saudi Arabia through various banks and from various Saudi businessmen.

[] What might be called tribute money to Al Qaida from countries that do not want the organization to establish cells in their own countries.

[] Funds from bin Laden's personal fortune.

Continued on next page

Osama (Usama) bin Laden and Islam, Continued

Bin Laden and Saudi Arabia

Even though the country where he grew up, Saudi Arabia, is extremely conservative, bin Laden finds that it is not conservative enough. He thinks the following reasons, which are contrary to his purist concept of Islam, are grounds for overthrowing the Saudi royal family:

[] Alliance of the ruling Saudi family with the West

[] Saudi Arabia's dependence on American and other foreign troops who came to the country to defend it during the Gulf War

[] Corruption of the Saudi regime

Appendix A: Short History of Terrorist Activity

Timeline The Activities of the Al Qaida

The Al Qaida has a long history of terrorist activity, which dates back to the early 1990s. Many of their earlier operations failed to achieve the desired effect; however, experience has enabled the organization to develop a more sophisticated approach.

[] Suspected in the bombing of night club in Bali
[] Suspected in September 11 plot to hijack planes and destroy the World Trade Center, Pentagon, and other unspecified targets (2001)
[] Plotted an unsuccessful effort to attack U.S. and Israel tourists during millennial celebrations (2000)
[] Implicated in bombings of U.S. embassies in Nairobi, Kenya and Dar es Salaam, Tanzania that killed over 300 (1998)
[] Accused by the U.S. of bomb attacks on American military personnel in Dhahran, Saudi Arabia (1996)
[] Linked to unsuccessful plot to assassinate President Clinton in the Philippines (1995)
[] Linked to unsuccessful plot to bomb multiple U.S. trans-Pacific flights (1995)
[] Linked to unsuccessful plot to bomb U.S. and Israeli embassies in Manila (1994)
[] Linked to unsuccessful plot to assassinate Pope John Paul II in Manila (1994)
[] Claims to have downed U.S. helicopters and killed U.S. servicemen in Somalia (1993)
[] Attempted bombing of U.S. troops in Aden, Yemen (1992)

Source: U.S. State Department

Appendix B: International Associations

Top Commanders

Al Qaeda is an international association of allied groups operating in many parts of the world. Its top commanders are

[] AYMAN AL ZAWAHRI, head of the Egyptian Jihad Islami, who is Bin Laden's senior deputy and heir apparent

[] JUMMA MAMANGANI, an Uzbek, who was recently appointed Al Qaeda chief of operations. He is former commander of the Moslem Army for the Liberation of Kyrgystan. Three key Afghan training camps, at Jalalabad, Farmada and Daronta, are under his command.

[] FATEH KAMEL, who leads the most militant cells of the extremist Algerian GIA. In the name of Al Qaida, he controls terrorist cells in the United States, Canada and Algeria.

[] MUHAMMED ATIF aka SUBHI ABU-SITTAH, who is nominal chief of staff of the network and its brightest military brain. He comes from the Egyptian Jihad Islami

[] IMAD MUGHNIYEH, the former Lebanese Hizballah hostage-taker and bomber, who is in charge of the combined terrorist campaign around the Middle East, the Persian Gulf, Europe and Israel.

Throughout the world

On top of the 3,500 hard core commanders and 110,000 Afghan and Pakistani troops, Al Qaeda retains another 6400 commanders in 12 centers: North America, Yemen, Saudi Arabia, Albania, Kosovo, Algeria, Chechenya, Tadjikistan and all the former Soviet republics of Central Asia, the Philippines, Egypt, Ethiopia and Somalia.

The numbers Bin Laden can muster differs from place to place. In North America, together with his closest ally, the Egyptian Jihad Islami, some 2,500 hard core fighting men; in Yemen, where his family originated before migrating to Saudi Arabia - 2000 directly. But Bin Laden has a special relationship with the commanders of the 20,000-man strong irregular "Moslem Liberation Army, which hold a monopoly of the arms trade of the Arabian Peninsula, the Horn of Africa, the Red Sea and East Africa. Its reach sometimes goes as far as Iran.

Bin Laden often serves as the MLA's clearing bank and ready bankroll for arms deal.

Continued on next page

Appendix B: International Associations, Continued

Saudi Arabia

In Saudi Arabia, where the privileged and affluent bin Laden clan lives, Osama commands some 200-250 hard-core commanders, but there are many more potential partisans among the disaffected tribes in the central and eastern provinces, especially the Nejd, as well as in the Saudi armed forces and national guard.

An intensive Al Qaeda recruitment effort in those Saudi forces could cause their collapse and drop in Bin Laden's lap their arsenals, with some of the most sophisticated hardware in use today.

Overseas Legions

According to conservative estimates, the millionaire-terrorist could most probably rally around the world roughly the same number of fighting men as those flocking to his flag in Afghanistan and Pakistan, namely an army of over 200,000 men. Many members of his overseas legions are available for both sustained and for one-time operations. In between, they simply go back to their normal pursuits and their homes in host countries. They are also available for terrorist operations on their home ground.

Appendix C: Other References

Resources

Besides the websites and URLs cited in the body of this handbook, the following resources will help you better understand Afghan culture and people:

[] *Encyclopedia of World Cultures* (particularly volume III), edited by Paul Hockings. Boston: G.K. Hall and Co., 1992. Excellent summaries of the different tribal and ethnic groups of Afghanistan and neighboring countries.

[] *The Europa World Year Book 2001* (volume I). London: Europa Publications, 2001. An excellent in-depth survey of Afghanistan.

[] Friedman, Norman. *Terrorism, Afghanistan, and America's NEW WAY OF WAR.* Annapolis: Naval Institute Press, 2003.

[] Hoffman, Frank G. (LtCol, USMCR, Ret.). "The U. S. Marine Corps in Review." *Proceedings,* May 2002: Vol. 128/5/, 84-90.

[] Hoffman, Frank G. (LtCol, USMCR, Ret.). "The U. S. Marine Corps in *Review." Proceedings,* May 2003: Vol. 129/5/, 96-102.

[] Marsden, Peter. *The Taliban: War, religion and the new order in Afghanistan.* New York: Zed Books Ltd., 1999. In-depth study of the Taliban including their origins and type of leadership, their religious "creed," implementation of religious beliefs, etc.

[] O'Balance, Edgar. *Afghan Wars 1839-1992: What Britain Gave Up and the Soviet Union Lost.* New York, Brassey's, 1993. Good account of the major conflicts in Afghanistan over the years: the Anglo Wars, the Soviet period, and civil wars.

[] Rubin, Burnett R. *The Search for Peace in Afghanistan: From Buffer State to Failed State.* New Haven: Yale University Press, 1995.

Made in the USA
Monee, IL
03 November 2026

67810974R00118